Spills of Nonfloating Oils
Risk and Response

Committee on Marine Transportation of Heavy Oils
Marine Board
Commission on Engineering and Technical Systems
National Research Council

NATIONAL ACADEMY PRESS
Washington, D.C.

NATIONAL ACADEMY PRESS • 2101 Constitution Avenue, N.W. • Washington, D.C. 20418

NOTICE: The project that is the subject of this report was approved by the Governing Board of the National Research Council, whose members are drawn from the councils of the National Academy of Sciences, the National Academy of Engineering, and the Institute of Medicine. The members of the committee responsible for the report were chosen for their special competencies and with regard for appropriate balance.

The National Academy of Sciences is a private, nonprofit, selfperpetuating society of distinguished scholars engaged in scientific and engineering research, dedicated to the furtherance of science and technology and to their use for the general welfare. Upon the authority of the charter granted to it by the Congress in 1863, the Academy has a mandate that requires it to advise the federal government on scientific and technical matters. Dr. Bruce M. Alberts is president of the National Academy of Sciences.

The National Academy of Engineering was established in 1964, under the charter of the National Academy of Sciences, as a parallel organization of outstanding engineers. It is autonomous in its administration and in the selection of its members, sharing with the National Academy of Sciences the responsibility for advising the federal government. The National Academy of Engineering also sponsors engineering programs aimed at meeting national needs, encourages education and research, and recognizes the superior achievements of engineers. Dr. William A. Wulf is president of the National Academy of Engineering.

The Institute of Medicine was established in 1970 by the National Academy of Sciences to secure the services of eminent members of appropriate professions in the examination of policy matters pertaining to the health of the public. The Institute acts under the responsibility given to the National Academy of Sciences by its congressional charter to be an adviser to the federal government and, upon its own initiative, to identify issues of medical care, research, and education. Dr. Kenneth I. Shine is president of the Institute of Medicine.

The National Research Council was organized by the National Academy of Sciences in 1916 to associate the broad community of science and technology with the Academy's purposes of furthering knowledge and advising the federal government. Functioning in accordance with general policies determined by the Academy, the Council has become the principal operating agency of both the National Academy of Sciences and the National Academy of Engineering in providing services to the government, the public, and the scientific and engineering communities. The Council is administered jointly by both Academies and the Institute of Medicine. Dr. Bruce M. Alberts and Dr. William A. Wulf are chairman and vice chairman, respectively, of the National Research Council.

This study was supported by the U.S. Coast Guard under Contract DTMA91-94-G-00003 between the National Academy of Sciences and the Maritime Administration of the U.S. Department of Transportation. Any opinions, findings, conclusions, or recommendations expressed in this publication are those of the author(s) and do not necessarily reflect the views of the organizations or agencies that provided support for the project.

International Standard Book Number 0-309-06590-9

Limited copies are available from: Marine Board, Commission on Engineering and Technical Systems, National Research Council, 2101 Constitution Avenue, N.W., Washington, D.C. 20418.

Additional copies of this report are available from National Academy Press, 2101 Constitution Avenue, N.W., Lockbox 285, Washington, D.C. 20055; (800) 624-6242 or (202) 334-3313 (in the Washington metropolitan area); Internet, http://www.nap.edu

Printed in the United States of America
Copyright© 1999 by the National Academy of Sciences. All rights reserved.

COMMITTEE ON MARINE TRANSPORTATION OF HEAVY OILS

MALCOLM L. SPAULDING, *chair,* University of Rhode Island, Narragansett
MALCOLM MacKINNON III, NAE, MSCL, Alexandria, Virginia
JACQUELINE MICHEL, Research Planning, Inc., Columbia, South Carolina
R. KEITH MICHEL, Herbert Engineering, San Francisco, California
JAMES L. O'BRIEN, O'Brien's Oil Pollution Service, Inc., Gretna, Louisiana
STEVEN L. PALMER, Florida Department of Environmental Protection, Tallahassee

Liaisons

PETER F. BONTADELLI, California Department of Fish and Game, Sacramento
MICHAEL C. CARTER/DANIEL LEUBECKER, Maritime Administration
BARBARA DAVIS, Environmental Protection Agency, Washington, D.C.
JERRY A. GALT, National Oceanic and Atmospheric Administration, Seattle, Washington
THOMAS HARRISON, United States Coast Guard, Washington, D.C.

National Research Council Staff

SUSAN GARBINI, Project Director
DONNA HENRY, Project Assistant
CAROL R. ARENBERG, Editor, Commission on Engineering and Technical Systems
DELPHINE D. GLAZE, Administrative Assistant

MARINE BOARD

JAMES M. COLEMAN, NAE, *chair*, Louisiana State University, Baton Rouge
JERRY A. ASPLAND, *vice chair*, California Maritime Academy, Vallejo
BERNHARD J. ABRAHAMSSON, University of Wisconsin, Superior
LARRY B. ATKINSON, Old Dominion University, Norfolk, Virginia
PETER F. BONTADELLI, California Department of Fish and Game, Sacramento
LILLIAN C. BORRONE, NAE, Port Authority of New York and New Jersey
BILIANA CICIN-SAIN, University of Delaware, Newark
SYLVIA A. EARLE, Deep Ocean Exploration and Research, Oakland, California
BILLY L. EDGE, Texas A&M University, College Station
JOHN W. FARRINGTON, Woods Hole Oceanographic Institution, Woods Hole, Massachusetts
MARTHA GRABOWSKI, LeMoyne College and Rensselaer Polytechnic Institute, Cazenovia, New York
R. KEITH MICHEL, Herbert Engineering, San Francisco, California
JEROME H. MILGRAM, NAE, Massachusetts Institute of Technology, Cambridge
JAMES D. MURFF, Exxon Production Research Company, Houston, Texas
STEVEN T. SCALZO, Foss Maritime Company, Seattle, Washington
MALCOLM L. SPAULDING, University of Rhode Island, Narragansett
ROD VULOVIC, Sea-Land Service, Charlotte, North Carolina
E.G. "SKIP" WARD, Shell Offshore, Houston, Texas

Staff

PETER JOHNSON, Acting Director
SUSAN GARBINI, Senior Staff Officer
DANA CAINES, Financial Associate
THERESA M. FISHER, Administrative Assistant
DONNA HENRY, Project Assistant

Preface

BACKGROUND

Maritime accidents that result in oil spills are high on the list of public environmental concerns. These spills are difficult to control and can contaminate the marine environment. When oil is spilled on the sea, it undergoes physical, chemical, and biological changes as it weathers and is degraded by bacteria. Most oil spill cleanup technologies, which have been developed for floating oils and the ensuing emulsions, are not very effective. For most spills, only about 10 to15 percent of the oil is recovered, and the best recovery rates are probably about 30 percent (OTA, 1990).

Some oils with a specific gravity greater than 1.0 (and some other oils in certain circumstances) may be neutrally buoyant or sink when spilled on water, depending on the salinity of the water. Federal rules governing oil spill contingency plans categorize petroleum cargoes according to their physical properties. Oils with a specific gravity of > 1.0, referred to as Group V oils, include some heavy fuel oils, asphalt products, and very heavy crude oils. Vessels and terminals that handle Group V oils are required to include responses to spills of Group V oils in their facility response plans.

The electric power generation industry often uses Group V oils because some Group V oil products are cheaper and have higher BTU content than other fuel oil products. Among these products are manufactured oils consisting of bitumen, water, and emulsifying agents. The presence of an emulsifying agent in the oil complicates the physical behavior of the oil if it is spilled into the water. Emulsified oils have been shown to sink initially to the level of their specific gravity and to surface later as the result of chemical changes caused by weathering.

Oils that sink to the bottom or remain suspended in the water column pose risks to certain resources that are not normally affected by floating oils. These resources include fish, shellfish, seagrasses, and other benthic (seabed) and water-column biota. Submerged oil may also cause episodic re-oiling of shorelines.

Although spills of Group V oils have been infrequent, there is some experience in responding to them and in cleaning them up. In most incidents in open water, oil in the water column is unrecoverable, and response operations are largely limited to locating and monitoring its movement. Where there is little or no current flow, suspended oil can sink and pool. In these cases, an effective response can be mounted, and most of the oil on the bottom can be recovered. Effective response (i.e., protecting the nearshore benthic communities) also means removing oil from the shoreline when and if it becomes stranded to keep it from being eroded and sinking in the nearshore tidal areas. Techniques that have been developed and demonstrated for recovering Group V oils following a spill include recovery of accumulations of oil on the seabed and vacuuming oily water for recovery in an oil-water separator. Other mechanical measures have also been investigated.

ORIGIN AND SCOPE OF THE STUDY

In the Coast Guard Authorization Act of 1996, the United States Coast Guard (USCG) was directed to assess the risk of spills for oils that may sink or be negatively buoyant, to examine and evaluate existing cleanup technologies, and to identify and appraise technological and financial barriers that could impede a prompt response to such spills. The USCG requested that the National Research Council (NRC) perform these tasks. In response to this request, the NRC established the Committee on the Marine Transportation of Heavy Oils under the auspices of the Marine Board.

The objectives of the study were: (1) to assess threats posed by the marine transportation of Group V oils by characterizing the trade of such oils and, in general terms, the resources at risk; (2) to assess the adequacy of cleanup technologies for spills of Group V oils and recommend research to develop new technologies and techniques, as appropriate; and (3) to identify barriers to effective responses to spills and recommend technological, financial, or management measures that would promote prompt and effective responses to spills of Group V oils. In discussions with the USCG and congressional staff, the committee clarified that the scope of study included the risk of oil spills and the capability of responding to them, although the environmental and health risks of spilled oil are not areas of the focus.

Committee members were selected with expertise in the following areas: the fate and effects of petroleum in water, habitats, and ecosystems; oil-spill response and cleanup technologies and operations; engineering systems analysis; tank vessel operations and port operations; environmental and regulatory issues; and relevant management and economic issues. Biographies of the committee members are provided in Appendix A.

Early in the committee's deliberations, it became clear that Group V oils, as defined by the USCG (oils with a specific gravity greater than 1.0), did not encompass all of the oils of concern. The drawbacks of using this narrow classification are that some Group V oils remain on the sea surface throughout the early response phase, while some lower density (e.g., Group IV) oils can be dispersed in the water column and sink to the seabed after weathering and interaction with sediments in the water column or after stranding onshore. The committee, therefore, decided to focus on the *behavior* of oil and use the term "nonfloating oils" as its operational definition. "Nonfloating oils" refers to oils that either initially or after weathering can be found in the water column or on the seabed; this definition includes oils that are suspended in the water column, sink to the seabed, or interact with sediments and are then deposited on the seabed or shoreline. The terms "sunken oils" or "submerged oils" are also used to describe oils that behave in this way.

The committee met four times during 1998 to gather information and discuss the issues of concern. At three of the meetings, presentations were made by a wide variety of individuals representing organizations in the transportation, spill response, environmental, scientific, and regulatory communities. A workshop was held in conjunction with the committee's second meeting to obtain information and to facilitate discussions of the issues. Leading experts in the marine transportation and spill response communities with expertise in the transport and response to spills of heavy or nonfloating oils participated in the workshop and panel discussions. Participants in the meetings and workshop are listed in Appendix B.

The committee's report is divided into five chapters. Chapter 1 focuses on the risk of spills of nonfloating oils and describes the traffic and trading patterns and recent history of heavy-oil spills, based on an analysis of available databases.

Chapter 2 describes the behavioral models for spills of nonfloating oils that can further an understanding of the fate and impact of these oils and be used to identify the resources at risk. This chapter also includes a comparative assessment of the environmental risks from spills of floating and nonfloating oils. Chapter 3 summarizes the technologies and techniques available for responding to spills of nonfloating oils. Subsections focus on modeling and information systems, spill tracking and mapping techniques, and containment and removal systems. Chapter 4 presents a discussion of the managerial, technological, and financial barriers to effective spill response. Chapter 5 presents the committee's findings, conclusions, and recommendations.

ACKNOWLEDGMENTS

The committee wishes to thank the many individuals who contributed their time and effort to this project by presenting material at committee meetings and workshops. Representatives of federal and state agencies, as well as private companies, provided invaluable assistance to the committee and the Marine Board staff.

This report has been reviewed by individuals chosen for their diverse perspectives and technical expertise, in accordance with procedures approved by the NRC's Report Review Committee. The purpose of this independent review is to provide candid and critical comments that will assist the authors and the NRC in making the published report as sound as possible and to ensure that the report meets institutional standards for objectivity, evidence, and responsiveness to the study charge. The content of the review comments and draft manuscript remains confidential to protect the integrity of the deliberative process. We wish to thank the following individuals for their participation in the review of this report:

John W. Farrington, Woods Hole Oceanographic Institution
Mervin F. Fingas, Environment Canada
Michael Herz, Marine Environmental Consultant
Donald S. Jensen, Jensen & Associates
Jerome H. Milgram, Massachusetts Institute of Technology
David Page, Bowdoin College
John Roberts, Coastal Towing

While the individuals listed above have provided many constructive comments and suggestions, responsibility for the final content of this report rests solely with the authoring committee and the NRC.

Contents

EXECUTIVE SUMMARY 1

1 TRANSPORTATION OF HEAVY OILS AND THE RISK OF SPILLS .. 9
   Definition of Terms, 9
   Overview of Quantitative Evaluation, 10
   Traffic and Trading Patterns, 10
   History of Spills, 14
   Projections of Spills, 18

2 BEHAVIORAL MODELS AND THE RESOURCES AT RISK 20
   Behavioral Models for Spills of Nonfloating Oils, 20
   Potential Effects of Nonfloating-Oil Spills, 30

3 TECHNOLOGIES AND TECHNIQUES 33
   Modeling and Information Systems, 33
   Tracking and Mapping Techniques, 37
   Containment and Recovery Methods, 40

4 BARRIERS TO EFFECTIVE RESPONSE 52
   Managerial Barriers, 52
   Technological Barriers, 53
   Financial Barriers, 54

5 **FINDINGS, CONCLUSIONS, AND RECOMMENDATIONS** 55
Findings, 55
Conclusions, 58
Recommendations, 59

REFERENCES 61

APPENDICES
A Biographical Sketches of Committee Members, 69
B Participants in the Workshop and Meetings, 72

Boxes, Figures, and Tables

BOXES

2-1   The *Nestucca* Spill, 25
2-2   The *Morris J. Berman* Spill, 26
2-3   The *Sansinena* Spill, 27

3-1   Oil-Spill Model, 34

FIGURES

1-1   Import/export and domestic movements of all crude oil and petroleum products in metric tons during calendar year 1996, 12
1-2   Import/export and domestic movement of crude oil and petroleum products in metric ton-miles during calendar year 1996, 13
1-3   Movements of petroleum by commodity in metric ton-miles during calendar years 1991 to 1996, 13
1-4   Movements of petroleum by tanker and tank barge in metric ton-miles during calendar years 1991 through 1996, 14
1-5   Volume of oil spilled from vessels in U.S. waters (1973 to 1996), 15
1-6   Geographical distribution of heavy-oil spills of 20 barrels or more from vessels in U.S. waters (1991–1996), 17

2-1 The relationship between water density and salinity at a temperature of 15°C, 21
2-2 Behavior of spilled nonfloating oils, 22
2-3a Oil-to-water density < 1.0; low sand interaction; majority of oil floats, 23
2-3b Oil-to-water density < 1.0; oil initially floats but sinks after stranding, 23
2-3c Oil-to-water density < 1.0; oil initially floats but sinks after mixing with sand in water, 23
2-3d Oil-to-water density > 1.0; low currents; majority of oil sinks, 24
2-3e Oil-to-water density > 1.0; high currents; oil disperses in water column, 24
2-4a Emulsified oil in freshwater; low currents; oil sinks, 29
2-4b Emulsified oil in freshwater; high currents; oil disperses and eventually sinks, 29
2-4c Emulsified oil in saltwater; high currents; oil initially disperses then coalesces into tarry slicks, 29

3-1 Decision tree based on oil density and water depth, 37
3-2 Decision tree for containment options for sunken oil, 46
3-3 Decision tree for recovery options for sunken oil, 47

TABLES

1-1 Movements of Petroleum by Tanker and Tank Barge during Calendar Years 1991 through 1996, 14
1-2 Oil Spills of 20 Barrels or More in U.S. Waters by Origin (1991 to 1996), 16
1-3 Heavy-Oil Spills of 20 Barrels or More in U.S. Waters by Origin (1991 to 1996), 17
1-4 Spill Rates for All Petroleum Cargoes in U.S. waters (1991 to 1996), 18
1-5 Spill Rates for Heavy Oil in U.S. Waters (1991 to 1996), 19

2-1 Relative Changes in the Resources at Risk from Spills of Nonfloating Oils Compared to Floating Oils, 31

3-1 Options for Tracking Oil Suspended in the Water Column, 42
3-2 Options for Mapping Oil Deposited on the Seabed, 44
3-3 Options for Containing Oil Suspended in the Water Column, 49
3-4 Options for Recovering Oil Deposited on the Seabed, 50

Executive Summary

In the Coast Guard Authorization Act of 1996, the United States Coast Guard (USCG) was directed to assess the risk of spills for oils that may sink or be negatively buoyant, to examine and evaluate existing cleanup technologies, and to identify and appraise technological and financial barriers that could impede a prompt response to such spills. The USCG requested that the National Research Council (NRC) perform these tasks. In response to this request, the NRC established the Committee on the Marine Transportation of Heavy Oils.

Early in the committee's deliberations, it became clear that the statutory definition of Group V oils (oils with a specific gravity greater than 1.0) did not include all of the oils of concern. The first problem with using this definition is that specific gravity is defined as the ratio of the density of oil to the density of freshwater at a fixed temperature. The density of seawater, however, is slightly higher than that of freshwater and increases as salt content increases. Therefore, Group V oils could have lower densities than those of the receiving seawater and float. The second problem is that an oil with a specific gravity of slightly less than 1.0 (e.g., a Group IV oil) might mix into the water column and sink to the seabed after weathering and interaction with sediments. The committee, therefore, decided to use the term "nonfloating oils" to include all of the oils of concern based on their behavior. Nonfloating oils move below the sea surface either because of their initial densities or because of changes in their densities as a result of weathering or interaction with sediments. These oils may be just below the water surface, suspended in the water column, or deposited on the seabed.

In order to carry out the assessment, the committee gathered the available data on the transportation and spills of Group V oils, as well as data on other oils that are known to sink or become suspended in the water column when weathered or mixed with sediment. The data were available for asphalt, coal tar, carbon black, bunker C, and No. 5 and No. 6 fuel oils, (i.e., so-called "heavy oils"). The committee used the USCG's (USCG) database on oil spills, refined with collaborative data from the Minerals Management Service (MMS), to develop estimates of the probability and mean size of oil spills. The U.S. Army Corps of Engineers (USACE) database on waterborne transportation of petroleum products and other cargoes over U.S. waters was used to assess the volumes of oil transported. The committee combined the spill statistics with the data on cargo tonnage to estimate historical rates on a barrel-per-ton-mile basis.

Historical spill rates must be modified for predictions of future spill rates because future rates will be influenced by fluctuations in traffic and trading patterns, as well as by changes in the ways vessels are designed and operated. The committee used the best available data, combined with its own collective judgment, to estimate the effects of these changes on the number and size of spills of nonfloating oils in the future.

Since 1991, the volume of oil spilled from vessels in U.S. waters has been reduced dramatically. Losses from tankers since 1990 have been less than one-tenth of the pre-1990 volume, and losses from barges have been less than one-third of the pre-1990 volume. From 1973 to 1990, there were 18 incidents involving spills of more than 25,000 barrels. Since 1991, there has not been a single spill of this magnitude for any category of oil. Nevertheless, very large spills will almost certainly occur some time in the future, although they are likely to be spills of crude oil rather than heavy oils, which tend to be transported in smaller volumes on barges and smaller tankers.

The USCG database includes descriptions of the substance spilled in each event. To estimate the frequency of spills of products with the potential to sink or become suspended in the water column after weathering or mixing with sediment, the committee summarized data for spills of more than 20 barrels for asphalt, coal tar, carbon black, bunker C, and No. 5 and No. 6 fuel oils. From 1991 to 1996, there was an average of 16 spills of these heavy oils per year, with an average volume of 785 barrels per spill. Tank barges were responsible for 28 percent of incidents and 80 percent of the volume of these spills of heavy oils. Most heavy-oil spills between 1991 and 1996 involved oils that were less dense than seawater, which only sink under unfavorable environmental conditions. The committee reviewed these heavy-oil spills with spill responders, who estimated that about 20 percent of these spills exhibited nonfloating behavior.

Most of the larger oil spills from land-based facilities were generally spills of crude oil or gasoline. The largest reported spill of heavy oil from a land-based facility between 1991 and 1996 was a spill of 929 barrels of No. 6 fuel oil into Pearl Harbor, Hawaii. By contrast, there were six tank-barge spills of more than 4,000 barrels involving heavy oil (either No. 6 fuel oil or slurry oil). The average volume of spills of heavy oil from barges was 2,254 barrels, and the largest was about 18,000 barrels. These spills were widely distributed geographically, with the highest frequency in the Gulf of Mexico.

Behavioral models have been developed for spills of nonfloating oils based on their physical and chemical properties. These descriptive, qualitative models predict how oils with densities near or above the density of the receiving water might behave. The models are based primarily on observations of oil spills. The committee described and assessed these models in terms of their effectiveness in predicting the behavior of nonfloating oils.

The environmental concerns associated with responses to spills of nonfloating oils are primarily related to water column and benthic (seabed) habitats. In most spills in open water, oil in the water column is unrecoverable, and response operations are limited to locating and monitoring its movement. However, if the suspended oil approaches shoreline habitats or nearshore benthic habitats in areas where current flow is minimal, the oil will sink and pool on the seabed. In these cases, an effective, but limited, response can be mounted, whereby a significant amount of oil can be removed from the seafloor. An effective response also includes removing oil from the shoreline, if and when it becomes stranded, to prevent its being eroded and sinking in nearshore tidal areas.

The behavior patterns of nonfloating oils can be complex, depending on the density of the oil, the density of the receiving water, and the physical characteristics of the spill site. Current technologies and techniques for locating, tracking, containing, and recovering spills of submerged oils include spill modeling and information systems, tracking and mapping techniques, and oil containment and recovery techniques. Chapter 3 focuses on the current state of practice and identifies systems that have been used or proposed for use in response to spills of nonfloating oils.

The containment and recovery of oil dispersed in the water column or deposited on the seabed is constrained by many factors, beginning with the difficulty of locating the oil and determining its condition. The success of current methods varies greatly but is usually limited because of the wide distribution of the oil and the fact that it is mixed with sediments and water. In general, available methods are most successful when the current speeds and wave conditions at the spill site are low (currents less than 10 cm/sec, wave heights less than 0.25 m), the oil is pumpable, the water is relatively shallow (water depths less than 10 m), and the sunken oil is concentrated in natural collection areas. The selection of methods for containment or recovery depends on the location and environmental conditions at the spill site, the characteristics of the oil and its state of weathering and interaction with sediments, and the equipment and logistical support available for the cleanup operation.

The committee identified a variety of barriers to responses to spills of nonfloating oils, including inadequate planning and training drills; lack of experience; lack of knowledge about transport, fate, and impact on the environment; the difficulty of locating and tracking oil suspended in the water column or deposited on the seabed; the limited technology options available for containment and recovery; and insufficient investment in research, development, testing, and evaluation of tracking, containment, and recovery systems.

FINDINGS

Finding 1. From 1991 to 1996, approximately 17 percent of the petroleum products transported over U.S. waters were heavy oils and heavy-oil products, such as residual fuel oils, coke, and asphalt. Approximately 44 percent was moved by barge and 56 percent by tanker.

Finding 2. From 1991 to 1996, approximately 23 percent of the petroleum products spilled in U.S. waters were heavy oils. In only 20 percent of these spills did a significant portion of the spilled products sink or become suspended in the water column. Most of the time, spills of heavy oil remained on the surface. The average number of spills of more than 20 barrels of heavy oil and asphalt was 16 per year, with an average volume of 785 barrels per spill. The committee projects that a 30 percent reduction in the number and volume of heavy-oil spills would have been realized if tankers and barges had all been double-hulled vessels.

Finding 3. In recent years, barges have had significantly higher spill rates than tankers. From 1991 to 1996, barges accounted for approximately 80 percent of the volume of heavy-oil spills, and the spill rate, expressed in terms of barrels-spilled-per-ton-mile, was more than 10 times higher for barges than for tankers. Although the reduction in spill volume from tank barges since 1990 has been significant (about one-third of pre-1990 volume), the reduction for tankers has been even more dramatic (about one-tenth of pre-1990 volume).

Finding 4. Specific gravity, as used in the regulatory definition of Group V oils, does not adequately characterize all oil types and weathering conditions that produce nonfloating oils. The committee was asked to address the issue of responses to Group V oil spills, defined by current regulations as oils with a specific gravity of greater than 1.0. However, the committee determined that the issue of concern is planning for and responding to oil spills in which most, or a significant quantity, of the spilled oil does not float. The committee, therefore, decided to use the term "nonfloating oils" to describe the oils of concern.

Finding 5. Nonfloating oils behave differently and have different environmental fates and effects than floating oils. The resources at greatest risk from spills of floating oils are those that use the water surface and the shoreline. Floating-oil spills seldom have significant impacts on water-column and benthic resources. In contrast, nonfloating-oil spills pose a substantial threat to water-column and benthic resources, particularly where significant amounts of oil have accumulated on the seafloor. Nonfloating oils tend to weather slowly and thus can affect resources for long periods of time and at great distances from the release site. However, the effects and behavior of nonfloating oil are poorly understood.

Finding 6. Although spill modeling and supporting information systems are well developed, they are not commonly used in response to nonfloating-oil spills because of limited environmental data and observations of oil suspended in the water or deposited on the seabed. Oil-spill models and supporting information systems are routinely used in contingency planning and spill responses. Sophisticated, user-friendly interfaces have been developed to take advantage of the latest advances in computer hardware and software. The current generation of models can rapidly incorporate environmental data from a variety of sources and include integrated geographic information systems. The models can also assimilate data on the most recently observed location of spilled oil and have improved forecasts of oil movements. They are not routinely used, however, in response to nonfloating-oil spills because of the lack of supporting data on the three-dimensional currents and concentrations of suspended sediments. Field data, such as oil concentrations in the water column and on the seabed, are also not generally available to validate or update models.

Finding 7. A substantial number of techniques and tools for tracking subsurface oil have been developed. Most of them, however, have not been used in response to actual oil spills. Many techniques are available for determining the location of oil both in the water column and on the seabed. These include visual observations, geophysical and acoustic methods, remote sensing, water-column and seabed sampling, *in situ* detectors, and nets and trawl sampling. The most direct and simplest methods, such as diver observations and direct sampling, are widely used, but they are labor intensive and slow. More sophisticated approaches, such as remote sensing, are limited to zones very near the sea surface because of technical constraints. Other advanced technologies, such as acoustic techniques, cannot differentiate between oil and water or between oiled sediments and underlying sediments. Many of the more sophisticated systems are prone to misuse and produce ambiguous data that are subject to misinterpretation. The performance of all but the simplest methods is undocumented either by field experiments or by use in spill responses.

Finding 8. Although many technologies are available for containing and recovering subsurface oil, few are effective, and most work only in very limited environmental conditions. Containment of oil suspended in the water column using silt curtains, pneumatic barriers, and nets and trawls is only effective in areas with very low currents and minimal wave activity. These conditions rarely exist at spill sites, particularly at sites in estuarine or coastal waters. The recovery of oil in the water column by trawls and nets is limited by the viscosity of the oil and net tow speeds.

The containment of oil on the seabed is typically ineffective, except at natural collection points (e.g., depressions and areas of convergence). The collection of oil on the seabed by manual methods, in natural collection areas and along the shoreline after beaching, is effective but labor intensive and slow. Manual methods are also limited by the depths at which diver-based operations can be carried out safely. Dredging techniques have rarely been used because of limited recovery rates, the large volumes of water and sediment generated, and the problems of storing, treating, and discharging co-produced materials.

Finding 9. The lack of knowledge and lack of experience, especially at the local level, in responding to spills of nonfloating oils is a significant barrier to effective response. The knowledge base and response capabilities for tracking, containing, and recovering nonfloating oils have not been adequately developed. Even at the national level, no system has been developed for sharing experiences or documenting the effectiveness and limitations of various options. With limited experience and a lack of proven, specialized systems, responders have found it difficult to adapt available equipment for responses to spills of nonfloating oils.

Finding 10. Planning for spills of nonfloating oils is inadequate at the local level. Existing area contingency plans do not include comprehensive sections on the risk of spills of nonfloating oils or how to respond to them. To date, planning has focused primarily on spills of floating oils. Inventories of equipment, lists of specialized services, assessments of the resources at risk, and protection priorities have not been developed by area committees for nonfloating oils. Nor have they identified the risks (e.g., transportation patterns, volumes, oil types), developed appropriate scenarios and response plans, or reviewed acceptable cleanup methods and end points. Existing plans have not been tested during drills or exercises to address deficiencies.

Finding 11. Funding levels for research, development, testing, and evaluation of spills of nonfloating oils are very low. The only active research programs currently under way either by government or industry groups are focused on emulsified fuel oils. Because the risk of spills of nonfloating oils is perceived as low relative to spills of floating oils, few research and development funds have been committed.

CONCLUSIONS

Conclusion 1. The tracking, containment, and recovery of spills of nonfloating oils pose challenging problems, principally because nonfloating oils suspended in the water column become mixed with large volumes of seawater and may interact with sediments in the water column or on the seabed. The ability to track, contain, and recover nonfloating oils is critically dependent on the physical and chemical properties of the oils and the water or the oils and the other materials dispersed in the water column or on the seabed. The differences in these characteristics are often quite small, and little technology is available for determining them.

Conclusion 2. Although many methods are available for tracking nonfloating oils, the simplest and most reliable are labor intensive and cover only limited areas. More sophisticated methods have severe technical limitations, require specialized equipment and highly skilled operators, or cannot distinguish oil from water or other materials dispersed in the water column. Engineered systems for containing oil in the water column or on the seabed are few and only work in environments with low currents and minimal waves. Natural containment in seabed depressions or in the lee of topographical or man-made structures on the seabed is effective for containing oils, but these are not always available in the vicinity of the spill.

Conclusion 3. The recovery of oil from the water column is very difficult because of the low concentration of dispersed oil; hence, recovery is rarely attempted. If oil collects on the seabed in natural containment areas, many options for effective recovery are available, although most of them are labor intensive and access to response equipment is a problem.

Conclusion 4. The volume and frequency of spills of nonfloating oils is significant (although smaller than for floating oils) and, therefore, should be an integral part of planning for spill responses, particularly in areas where nonfloating oils are regularly transported. Transport by tank barges raises particular concerns, given the relatively high spill rates from these vessels. The risks of potential harm to water-column and benthic resources from nonfloating oils have not been adequately addressed in the contingency plans for individual facilities or geographic areas.

Conclusion 5. Inland barges are subject to greater risks of spills than tankers and coastal barges; consequently, spill rates for barges are likely to be higher than for tankers. However, the large difference between the overall spill rates, as well as the decreasing number of spills from tankers in recent years (post-OPA 90), raises concerns regarding the performance of barges.

RECOMMENDATIONS

The recommendations below are intended to improve the capability of the spill response community to respond to spills of nonfloating oils.

Recommendation 1. The U.S. Coast Guard should direct area planning committees to assess the risk of spills of nonfloating oils (i.e., oils that may be dispersed in the water column or ultimately sink to the seabed) to determine the resources at risk. In areas with significant environmental resources risk, area planning committees should develop response plans that include consultation and coordination protocols and should obtain pre-approvals and authorizations to facilitate responses to spills. Stakeholder groups should be educated about the impact and methods available for tracking, containing, and recovering oil suspended in the water column or on the seabed. Area committees in locations where there is a high risk of spills of nonfloating oils should include at least one scenario for responding to a nonfloating-oil spill in their training or drill programs.

Recommendation 2. The U.S. Coast Guard should improve its knowledge base, education, and training for responding to spills of nonfloating oils by including a scenario involving a spill of nonfloating oils in oil-spill response drills, by establishing a knowledge base and scientific support teams to respond to these types of spills, and by disseminating this knowledge to the federal spill-response coordinators and area planning committees as part of ongoing training programs. The information would help area planners assess the requirements for responding to nonfloating-oil spills.

Recommendation 3. The U.S. Coast Guard should support the development and implementation of an evaluation program for tracking oil in the water column and on the seabed, as well as containment and recovery techniques for use on the seabed. The findings of these evaluations should be documented and distributed to the environmental response community to improve response plans for spills of nonfloating oils.

Recommendation 4. Tests of area contingency plans and industry response plans for responses to spills of nonfloating oils should be required parts of training and drill programs.

Recommendation 5. The U.S. Coast Guard should monitor spill rates from tank barges to ascertain whether current regulatory requirements and voluntary programs will reduce the frequency and volume of spill incidents. If not, the Coast Guard should consider initiating regulatory changes.

1

Transportation of Heavy Oils and the Risk of Spills

An assessment of the risk of spills involves evaluating the frequency and consequences of accidents. A formal assessment of consequences should be based on a wide range of factors, including loss of life, financial loss, and short-term and long-term environmental impacts. In this chapter, the quantity of oil spilled is considered. Between 1991 and 1996, domestic tanker operations were responsible for nearly 75 percent of the ton-miles of petroleum movements. The major component is the coastal movement of Alaskan North Slope oil to U.S. ports on the West Coast.

DEFINITION OF TERMS

Group V oils are defined as persistent oils with a specific gravity of greater than 1.0 (Federal Register, 1996). *Heavy oil* is the term used by the response community to describe dense, viscous oils with the following general characteristics: low volatility (flash point higher than 65°C), very little loss by evaporation, and a viscous to semisolid consistency (NOAA and API, 1995). Examples of heavy oils include Venezuela crude, San Joaquin Valley crude, Bunker crude, and No. 6 fuel oil. The term heavy oil, in this chapter, also refers to residual oils (No. 5 and No. 6 fuel oil, Bunker C, and slurry oil), asphalt, coal tar, coke, carbon black, and pitch.

The term *nonfloating oil* is used to describe all oils that do not float on water, including oils that are denser than the receiving waters and either sink immediately or mix into the water column and move with the water as suspended oil; as well as the portion of oil that is initially buoyant but sinks after interacting with sand. The committee chose not to use the term *sinking oil*, which implies that the oil sinks directly to the bottom, because it would not include all of the types of oil and spill conditions of concern in this report. *Emulsified fuels* (anthropogenic fuels manufactured by mixing water with liquid oils or solid hydrocarbon products), for example, often contain a surfactant to stabilize the emulsion and can be dispersed in the water column.

OVERVIEW OF QUANTITATIVE EVALUATION

The historical frequency of oil spills in general and heavy-oil spills in particular can be estimated from spill statistics. The committee used the U.S. Coast Guard (USCG) database on oil spills, refined with collaborative data from the Mineral Management Service (MMS), to estimate the probability and mean size of oil spills. The U.S. Army Corps of Engineers (USACE) database on the waterborne transportation of petroleum products and other cargoes in U.S. waters was used to assess the volume of oil transported. By combining the statistics on spills with the data on cargo tonnage, the committee was able to estimate historical spill rates on a barrel-per-ton-mile. Because future spill rates may be influenced by fluctuations in traffic and trading patterns, as well as changes in vessel design and operation, these estimates should be reevaluated to predict future rates. The committee has combined the best available data with its own collective judgment in these estimates. It should be noted that in only 20 percent of spills of heavy oil does a significant portion of the spilled oil sink or become suspended in the water column.

TRAFFIC AND TRADING PATTERNS

The USACE (1998a, 1998b) compiles detailed statistics on U.S. waterborne commerce, both foreign (imports and exports) and domestic (trade between U.S. ports). Domestic movements are further subdivided into coastal trade (involving carriage over the ocean) and internal trade (solely on inland waterways).

Figure 1-1 summarizes the data for all movements of crude oils and petroleum products during 1996 (the most recent data available). The USACE data are separated into 19 commodity codes, but for the sake of simplicity, the committee combined some categories (e.g., gasoline and kerosene) into seven categories (*crude oil*; *residual fuel oil*; *coke, tar, pitch, asphalt*; *gasoline, kerosene*; *distillate fuel oil*; *naptha, solvents*; and *lubrication, grease, wax*). The substances in the *residual fuel oil* and *coke, tar, pitch, and asphalt* categories are heavy oils (i.e., they are either heavier than water or have the potential of sinking or becoming suspended in the water column upon weathering).

Crude oil accounted for 56 percent of the total tonnage of the petroleum commodities shipped in 1996; international trade accounted for 76 percent. The largest component of the domestic trade in crude oil was the coastal movement of

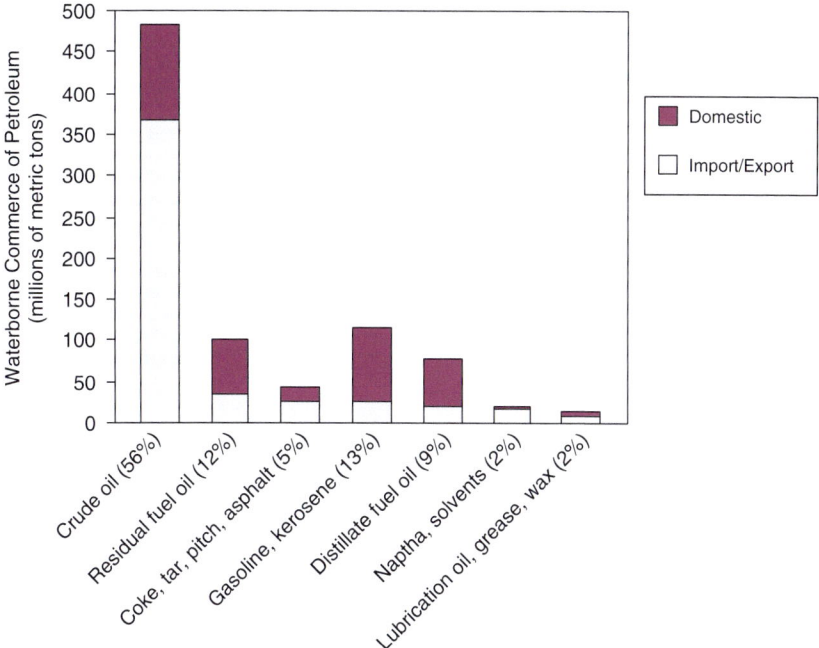

FIGURE 1-1 Import/export and domestic movements of all crude oil and petroleum products in metric tons during calendar year 1996. Source: USACE, 1998a.

Alaskan North Slope oil to U.S. ports on the West Coast. Internal trade (solely on inland waterways) accounted for less than 5 percent of the total. Nearly all of the international tonnage and 70 percent of the domestic tonnage was shipped in tankers. The very heavy crude oils produced in the United States (e.g., California crudes, such as San Joaquin and Santa Maria) were transported primarily through overland pipelines. Some very heavy crude oils were also imported (e.g., from Venezuela and Mexico), but these are believed to comprise only a small fraction of the total volume of imported crude oil.

Residual fuel oils are represented by a single code in the USACE database, which includes Nos. 5 and 6 fuel oils and slurry oils. Residual fuel oils accounted for 12 percent of the total tonnage of petroleum products shipped in 1996; coke, tar, pitch, and asphalt accounted for another 5 percent of the total. The combined total for heavy oils was, therefore, 17 percent of the total movement of all oil and petroleum products. Approximately 90 percent of the domestic waterborne trade of these heavy oils was transported by barge (whereas more than 90 percent of the international trade was transported by tanker). Overall, therefore, about 44 percent of heavy oils was transported by barge and 56 percent by tanker.

Group V oils are transported along the Gulf Coast from Corpus Christi to

New Orleans, from the Gulf Coast upriver to the St. Louis area, and along the Ohio River to ports further inland. Some Group V oils are also produced in St. Paul, Minnesota, and transported down the upper Mississippi River and up the Ohio River. Heavy residual oils are transported to power generating facilities through the inland waterways and along the East Coast and Gulf Coast and are exported from California to the Far East. Asphalt is moved in tankers and tank barges along the coasts (primarily along the Gulf and East coasts) as both imports and domestic cargoes, and in barges along the inland waterways. Some very heavy crude oils (e.g., Venezuela Boscan crude) are imported to East and Gulf coast refineries. Carbon black feedstock moved among refineries on the Gulf Coast and was exported from California. Bunkering fuels for ships (typically No. 6 fuel oil) are moved intra-harbor on barges. Most large commercial ships (including containerships, dry bulk carriers, tankers, cruise ships, as well as some tugboats) use these heavy oils as fuel, although these oils are not included in the statistics on the waterborne commerce of petroleum.

In Figure 1-2, the movement of crude oils and petroleum products for calendar year 1996 are shown in *metric ton-miles*. The *domestic ton-miles* are calculated by multiplying the metric tons of cargo being transported by the number of miles actually moved on the water. The average length of a domestic voyage was about 900 miles. For imports and exports, a constant of 100 miles per voyage was

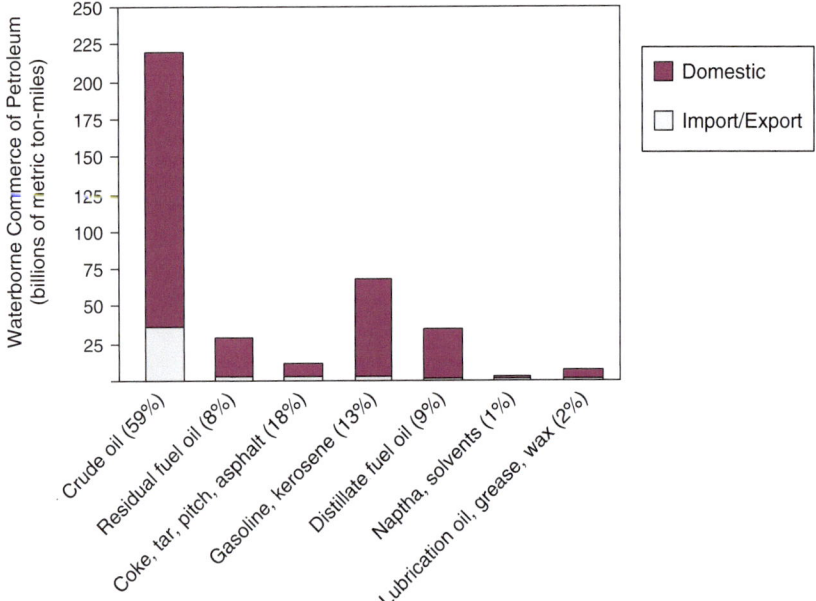

FIGURE 1-2 Import/export and domestic movements of crude oil and petroleum products in metric ton-miles during calendar year 1996. Source: USACE, 1998b.

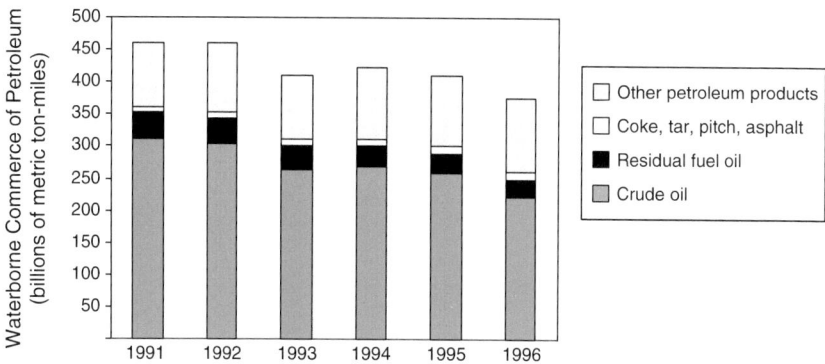

FIGURE 1-3 Movements of petroleum by commodity in metric-ton miles during calendar years 1991 to 1996. Source: USACE, 1998b.

assumed to account for the exposure of the vessel when transiting coastal waters and navigating U.S. waterways and channels. Movements of residual fuels comprised only 8 percent of the total.

The U.S. waterborne commerce of petroleum gradually decreased from 1991 to 1996 (Figure 1-3), primarily as a result of cutbacks in the coastal tanker trade of crude oil. During this period, the movement of residual fuel oils declined by 45 percent, due partly to improvements in the refining process, which produces less residual oil per barrel of crude oil refined. The movement of coke, tar, pitch, and asphalt, however, increased by 47 percent. Preliminary USACE figures for 1997 indicate that the domestic trade for coke, tar, pitch, and asphalt was up nearly 70 percent compared to 1996.

Movements of petroleum by tanker and tank barge are summarized for calendar years 1991 through 1996 in Figure 1-4 and Table 1-1. Figure 1-4 shows that domestic barge traffic remained relatively constant during the period. Tanker import and export traffic increased by about 5 percent per year, reflecting increases in imports of crude oil; the tanker domestic traffic declined by about 7 percent per year.

The U.S. Department of Energy (DOE, 1998) estimates that the percentage of petroleum consumption met by imports will increase from 49 percent in 1997 to 65 percent in 2020. This increase is partially a reflection of anticipated reductions in domestic production as oil reserves are depleted and a projected 1.1 percent per year increase in domestic energy consumption. The higher demand will probably be met through increased imports of long-haul crude oil (NRC, 1998). Future trends in the movements of heavy residual oils and asphalt are more difficult to quantify. The committee heard several presentations on the interest of some utility companies in using emulsified fuels (e.g., Orimulsion™) to generate power. Emulsified fuels do not float on water and are included in the

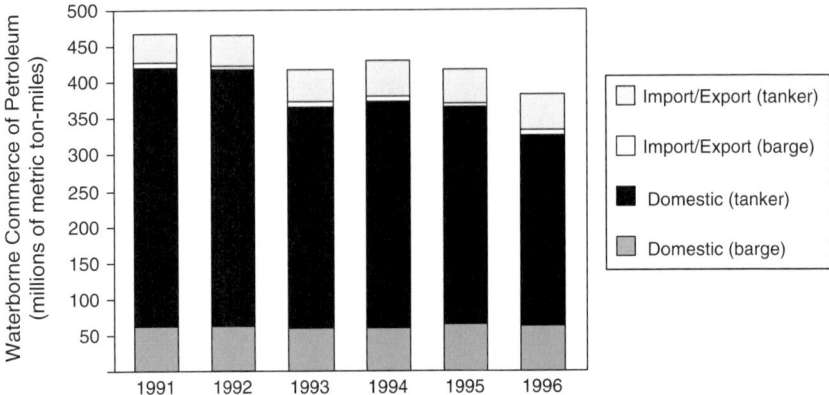

FIGURE 1-4 Movements of petroleum by tanker and tank barge in metric ton-miles during calendar years 1991 through 1996. Source: USACE, 1998b.

definition of nonfloating oils. Environmental groups responding to a proposal to burn a Venezuelan emulsified fuel for power generation in Manatee County, Florida, expressed concerns about cleaning up emulsified fuel spills once the oil had dispersed into the water column (Rains, 1998). Another concern was air quality because these fuels tend to be high in sulfur and other contaminants. At this point, it is difficult to project the consumption of emulsified fuels in the United States.

HISTORY OF SPILLS

The historical data on oil spills in U.S. navigable waters was derived from both the USCG and MMS databases. The USCG database includes reported oil spills of all sizes in U.S. navigable waters. Although these data are

TABLE 1-1 Movements of Petroleum by Tanker and Tank Barge during Calendar Years 1991 through 1996

	U.S. Waterborne Traffic in Metric Ton-Miles (× 1 billion)			
	1991-1996 Mean		1996 Totals	
	Barge	Tanker	Barge	Tanker
Crude Oil	4.9	266.8	4.9	215.4
Residual Fuel Oil	12.1	23.5	12.7	16.8
Coke, Tar, Pitch, Asphalt	7.4	2.5	8.7	3.6
Other Petroleum Products	37.8	60.5	27.8	76.1
Totals	62.2	361.3	64.1	311.9

comprehensive, they have not been uniformly maintained over the years. The MMS database has been consistently maintained but only covers spills of more than 1,000 barrels from tankers and tank barges. By comparing the USCG data with the MMS data, the committee has modified the USCG data, as necessary.

Since 1991, there has been a dramatic reduction in the volume of oil spilled from vessels in U.S. waters (Figure 1-5). Losses from tankers since 1990 are less than one-tenth the volume of pre-1990 losses, and losses from barges are less than one-third the volume of pre-1990 losses. From 1973 to 1990, there were 18 spills of more than 25,000 barrels each. Since 1991, there has not been a single spill of this magnitude. This statistic may be fortuitous, however, and a very large spill is likely to occur in the future. Large future spills are likely to involve crude oil rather than heavy oil, however, because most heavy oils and asphalt are carried on barges and smaller tankers.

In light of the huge decrease in the number of oil spills since 1990, the committee based its projections on the 1991 to 1996 data. Because of inconsistencies in the data for small spills, the committee limited its analysis to spills of more than 20 barrels, which account for more than 98 percent of the spills in this period.

Spills in the USCG database are divided into two general categories based on their origin: vessels and facilities. Facilities include pipelines, ground

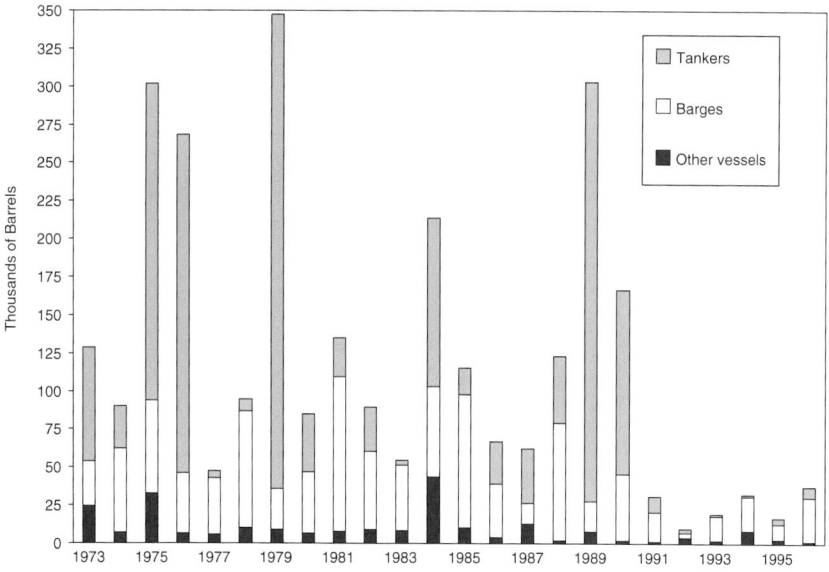

FIGURE 1-5 Volume of oil spilled from vessels in U.S. waters (1973 to 1996). Sources: USCG, 1998; MMS, 1998.

transportation onshore facilities (e.g., shoreside structures, such as terminals, refineries, and storage tanks), and offshore facilities (e.g., drilling rigs and production platforms). Vessels are subdivided into tankers, tank barges, and other vessels (ships not engaged in the transport of petroleum). The number and volumes of spills are summarized in Table 1-2. Although tankers were the primary source of marine oil spills prior to 1990, facilities have been responsible for a majority of the incidents and most of the total spill volume since then. Pipelines have been the source of more than 50 percent of the spill volume from facilities, and tank barges for more than 75 percent of the spill volume from vessels.

The USCG database provides a description of the substance spilled in each event. Table 1-3 summarizes data for all spills of more than 20 barrels of nonfloating oils (i.e., products with the potential to sink or become suspended in the water column when weathered or mixed with sediment). These products include asphalt, coal tar, carbon black, bunker C, and Nos. 5 and 6 fuel oils. Spills of nonfloating oils constitute about 23 percent of the total volume of oil spilled. From 1991 to 1996, the average number of spills of nonfloating oils was 16 per year, with an average volume of 785 barrels per spill. Tank barges were responsible for 28 percent of incidents and 80 percent of the total spill volume.

Releases of 20 barrels or more from facilities were generally spills of floating oils (either crude oil or gasoline). The largest reported spills of heavy oils from a facility was a spill of 929 barrels of No. 6 fuel oil in Pearl Harbor, Hawaii. In contrast, there were six spills from tank barges of more than 4,000 barrels each, all of them of heavy oils (either No. 6 fuel oil or slurry oil). The average volume of heavy-oil spills from barges was 2,254 barrels, and the largest spill during this period was about 18,000 barrels. Spills were widely distributed geographically (Figure 1-6), with the highest frequency from vessels in the Gulf of Mexico. Some of the oils categorized as heavy oils in the USCG and MMS databases are less dense than seawater and will remain afloat under certain environmental conditions. To determine the frequency of nonfloating-oil spills, the committee examined heavy-oil spills of more than 20 barrels (a total of 93 spills)

TABLE 1-2 Oil Spills of 20 Barrels or More in U.S. Waters by Origin (1991 to 1996)

	No. of Incidents		Total Spill Volume (barrels)		Average Spill Volume (barrels)
Tankers	47	(8%)	26,508	(8%)	564
Tank barges	100	(17%)	100,785	(32%)	1,008
Other vessels	44	(7%)	11,474	(4%)	261
Facilities	415	(68%)	173,945	(56%)	419
1991 to 1996 totals	606		312,713		
Average per year	101		52,119		

TABLE 1-3 Heavy-Oil Spills of 20 Barrels or More in U.S. Waters by Origin (1991 to 1996)

	No. of Incidents		Total Spill Volume (barrels)		Average Spill Volume (barrels)
Tankers	17	(18%)	6,442	(9%)	379
Tank barges	26	(28%)	58,591	(80%)	2,254
Other vessels	22	(24%)	3,877	(5%)	176
Facilities	28	(30%)	4,083	(6%)	146
1991 to 1996 totals	93		729,913		
Average per year	16		12,166		

to identify the spills in which a significant fraction of the oil did not float. These spills accounted for about 20 percent of the heavy-oil spills and about 50 percent of the volume of heavy oil spilled during this period. The relatively high volume of nonfloating-oil spills, as compared to the relatively low number of nonfloating-oil spills (20 percent), is attributable to a few large heavy-oil spills during the period. One spill in particular, the *Morris J. Berman* spill of nearly 18,000 barrels of heavy oil in 1994, strongly influenced the statistics.

The committee could not explain why the average volume of nonfloating-oil spills should differ from the average volume of heavy-oil spills and considers the high volume of nonfloating-oil spills to be an anomaly caused by the limited statistics. Assuming that nonfloating-oil spills comprise 20 percent of the heavy-

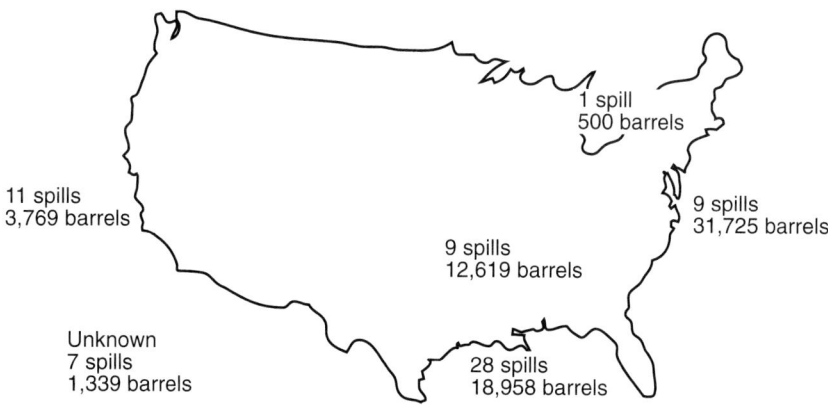

FIGURE 1-6 Geographical distribution of heavy-oil spills of 20 barrels or more from vessels in U.S. waters (1991 to 1996).

oil spills by number, the committee estimates that the average per year (20 percent of 16) will be three or four nonfloating-oil spills. Assuming that the average volume of nonfloating-oil spills is the same as for heavy-oil spills (i.e., 785 barrels per spill), the projected volume will be about 2,500 barrels per year.

PROJECTIONS OF SPILLS

To assess the risk of heavy-oil spills from vessels, the committee used ton-miles as a measure of exposure and the quantity of oil spilled as a measure of the consequences of accidents. Based on this approach, the spill rate is defined as the ratio of the historic volume of oil spilled to the historic movements in ton-miles and is expressed as barrels spilled per billion ton-miles. Tankers and tank barges were responsible for 89 percent of the heavy-oil spills from 1991 to 1996. The spill rates for all petroleum cargoes and for heavy-oil cargoes are presented in Tables 1-4 and 1-5, respectively. Barges had higher spill rates for all petroleum cargo than tankers during this period. The spill rates for heavy oil carried by barges were higher by a factor of two than the spill rates for all petroleum cargoes. The spill rates in Table 1-5 are for heavy oils, some of which remain afloat under certain environmental conditions. Only about 20 percent of the heavy oil spilled is expected to exhibit nonfloating behavior.

The volume of future spills will be affected by changes in the design and operation of tankers and barges. Decreases in both the number and volume of oil spills are expected as the fleet completes the transition to double-hull construction (NRC, 1998).

The spill statistics suggest that the barge industry has lagged behind the tanker industry in improving operations since the Oil Pollution Act of 1990 (OPA 90) was enacted. Major barge accidents from 1991 to 1996 had a variety of causes, including structural failure, capsizing, allisions, collisions, and groundings. The barge industry has instituted a number of voluntary programs to improve its environmental performance and safety record. These include the American Waterways Operators Responsible Carrier Program and partnerships with the USCG.

TABLE 1-4 Spill Rates for All Petroleum Cargoes in U.S. Waters (1991 to 1996)

	Movement of Petroleum (billions of metric ton-miles per year)	Oil Spill Volume (barrels per year)	Spill Rate (barrels spilled per billion metric ton-miles)
Tanker	361.3	4,418	12
Barge	62.2	16,798	270

TABLE 1-5 Spill Rates for Heavy Oil in U.S. Waters (1991 to 1996)

	Movement of Heavy Oil (billions of metric ton-miles per year)	Oil Spill Volume (barrels per year)	Spill Rate (barrels spilled per billion metric ton-miles)
Tanker	26.1	1,074	41
Barge	19.6	9,765	499

From 1991 to 1996, the percentage of tonnage carried in double-hull vessels was approximately 13 percent for tankers and 60 percent for barges. Theoretical comparisons with single-hull vessels (NRC, 1998) indicate that double-hulled tankers and tank barges will be involved in four to six times fewer spills. If all vessels trading from 1991 to 1996 had been double-hull, the number and volume of heavy oil spills could have been reduced by about 30 percent. In accordance with the provisions of OPA 90, the transition to double-hull vessels will be completed by January 1, 2015.

Total cargo movements in U.S. waters have increased at an average rate of 2 percent per year for the past 10 years. Further growth will tend to increase the number of spills from bunkers on freighters and other commercial vessels, and a move is under way to protectively locate bunker tanks on larger tankers and a few large container ships, which should lead to a reduction in the spillage of fuel oil.

2

Behavioral Models and the Resources at Risk

BEHAVIORAL MODELS FOR SPILLS OF NONFLOATING OILS

Based on an understanding of the physical and chemical properties of nonfloating oils (mostly from observations of past spills), Behavioral models have been developed (Michel et al., 1995). These models are descriptive, qualitative predictions of how oils with a density near or higher than the density of the receiving water might behave. The key factors that determine the behavior of spilled nonfloating oils are: water density, current speed, and the potential for interaction with sand.

Water Density

If the ratio of the density of oil to the density of the receiving water is greater than 1.0, the oil will not float. If it less than 1.0, the oil will float. If it is within a few percent of 1.0, then the oil is much more likely to become submerged by wave action. Figure 2-1 shows the relationship between the density and salinity of the water for a fixed temperature. The density is also shown in terms of the API (American Petroleum Institute) gravity. Oils with higher densities than the receiving water (above the line) will sink; oils with lower densities that the receiving water (below the line) will initially float.

Current Speed

If current speeds are greater than 0.1 m/s, nonfloating oils will be suspended in the water column. If the currents are very slow, oils heavier than the receiving water will sink to the bottom (Nielsen, 1992).

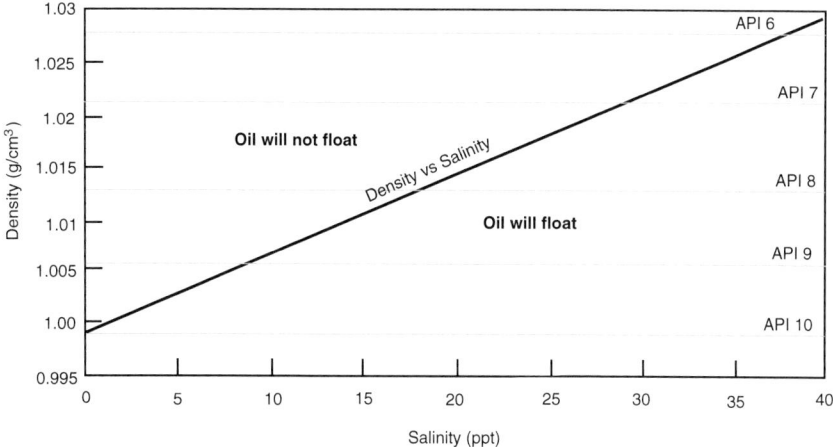

FIGURE 2-1 The relationship between water density and salinity at a temperature of 15°C. The density is also shown in API gravity units (right vertical axis).

Potential for Interaction with Sand

When floating oil is mixed with 2 to 3 percent sand, it becomes heavier than water and sinks (Michel and Galt, 1995). The density of sand grains is much higher than the density of silt or clay particles. Figure 2-2 is a schematic illustration showing the relationships among these factors and how they affect the short-term behavior of nonfloating oils. The density of oil relative to the receiving water is important only in determining whether the oil will initially float. Significant currents can keep heavier-than-water oil suspended in the water column. Any oil still on the surface or suspended in the water column will still sink if it mixes with sand also suspended in the water column. The models in Figures 2-3 and 2-4 illustrate combinations of factors that influence the behavior of nonfloating oils.

Oil Lighter than Water, Low Sand Interaction

If the oil-to-water density ratio is less than 1.0, the oil will initially float. At 15°C, oils with an API gravity above 6.5 (Figure 2-1) will still be lighter than seawater with a salinity of 35 parts per thousand. These oils will float, at first in contiguous slicks that may quickly (often within a few kilometers) break up into widely scattered fields of large mats and tar balls that can spread over large distances and become reconcentrated again in convergence zones (Figure 2-3a).

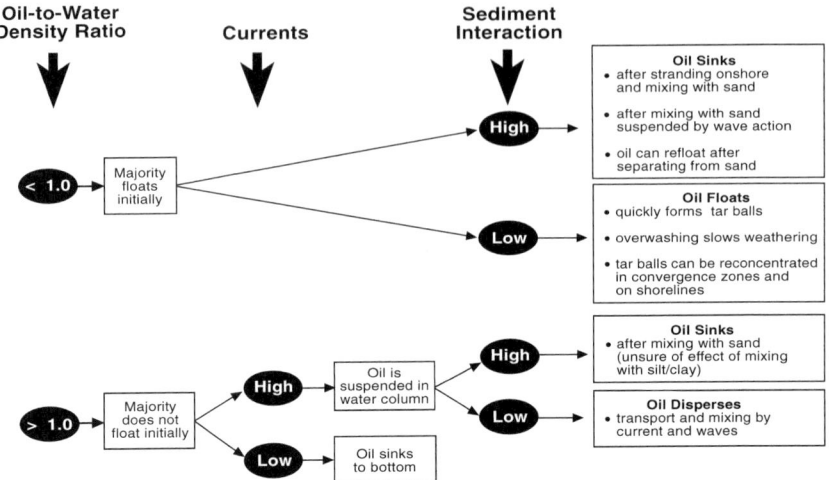

FIGURE 2-2 Behavior of spilled nonfloating oils.

Because of the higher viscosities of heavy oils, the tar balls are more persistent than for spills of light and medium oils. More important, however, as the density of the oil approaches that of the water, these tar balls tend to become "overwashed" by wave action making them very difficult to track and slowing most weathering processes (e.g., evaporation or formation of a "skin") (Lee et al., 1989). Furthermore, if oil emulsifies, the emulsion can contain 50 to 80 percent water making the density of the oil even closer to the density of the water. Evaporation of emulsified oils is slow, and, unless they interact with sediment, they will remain floating. When tar balls are eventually stranded, sometimes hundreds of kilometers away from the original spill site, the oil can still be relatively fresh and have a significant impact on the water surface and shoreline resources (see Box 2-1). Because the oil still floats, this type of spill is not considered further in this report. After the evaporative loss of the lighter fraction, particularly of the cutter stock in bunker fuels, the remainder might sink, but this has been observed at only one spill (Lee et al., 1992; Michel and Galt, 1995).

Oil Lighter than Water, High Sand Interaction

Spilled oil that is lighter than the receiving water can still sink, either by becoming stranded on sand beaches or by mixing with sand in the surf zone. In several spills, such as the IXTOC I (Gundlach et al., 1981), *Alvenus* (Alejandro and Buri, 1987), and *Haven* (Martinelli et al., 1995), heavy oils floated initially

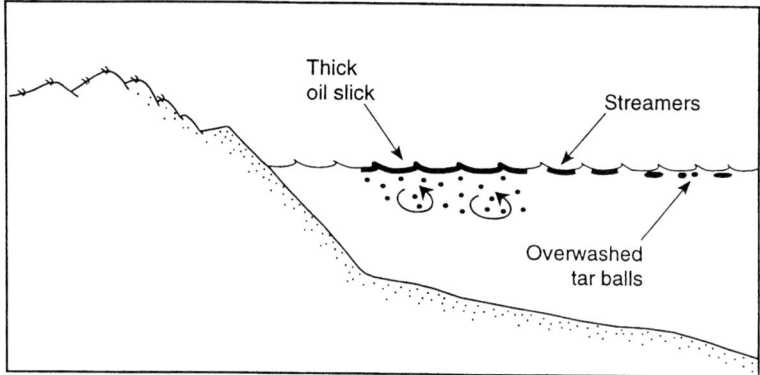

FIGURE 2-3a Oil-to-water density < 1.0; low sand interaction; majority of oil floats.

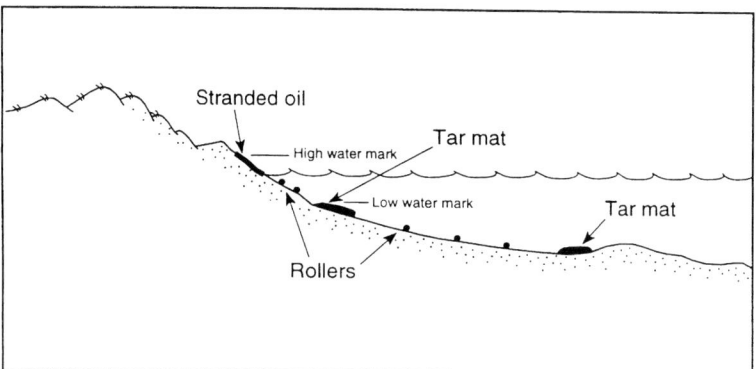

FIGURE 2-3b Oil-to-water density < 1.0; oil initially floats but sinks after stranding.

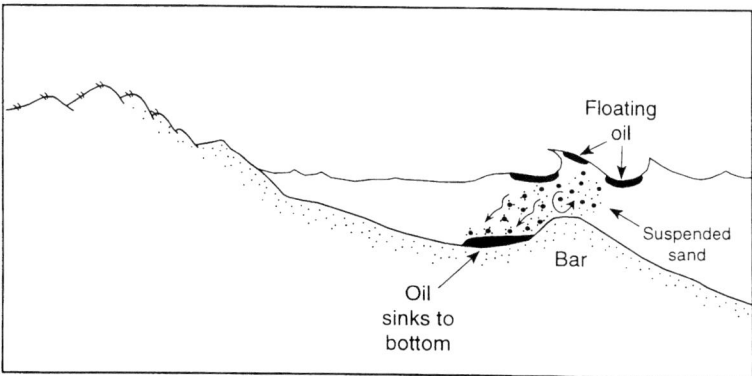

FIGURE 2-3c Oil-to-water density < 1.0; oil initially floats but sinks after mixing with sand in water.

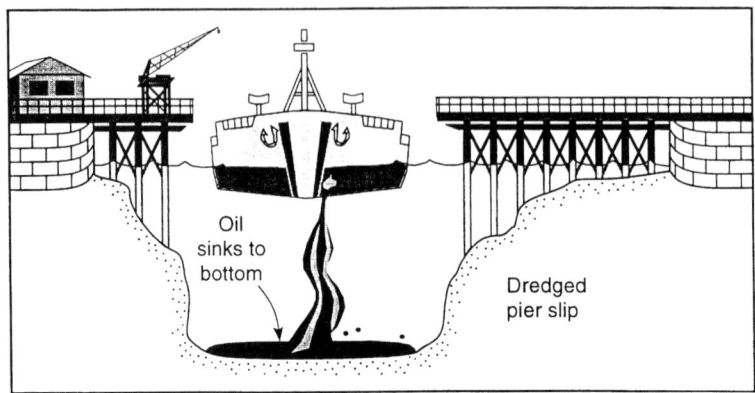

FIGURE 2-3d Oil-to-water density > 1.0; low currents; majority of oil sinks.

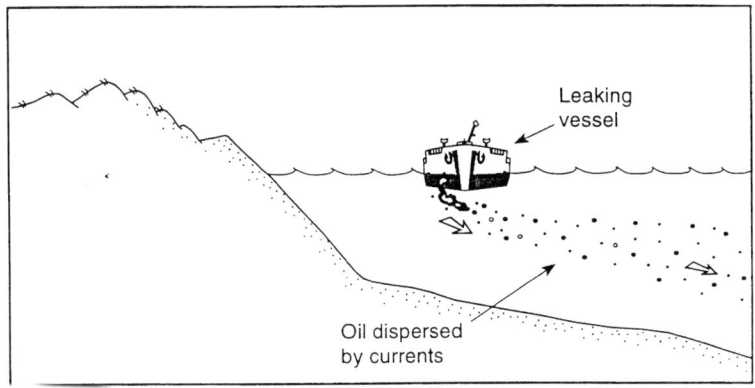

FIGURE 2-3e Oil-to-water density > 1.0; high currents; oil disperses in water column.

and became stranded on sand beaches but then were eroded from the beaches and sank, forming tar mats between nearshore bars. In these cases, the oil was too viscous to penetrate the sand; instead, the sand coated the oil layers and/or mixed with the viscous oil as it was eroded from the beaches by wave action. The oil/sand mixture contained only a few percent sand and was deposited at the toe of the beach just offshore (Figure 2-3b). The distribution of sunken oil/sand tar mats was highly variable, ranging from thick, continuous deposits tens of meters long to small widely scattered tar balls. If there was current activity, especially generated by waves breaking on the shore, the oil/sand mixture formed cigar-shaped "rollers" that were scattered on the bottom or accumulated into mats in the

**BOX 2-1
The *Nestucca* Spill**

The *Nestucca* spill in December 1988 released 5,500 barrels of heavy marine fuel oil with an API gravity of 12.1 three kilometers off Grays Harbor, Washington. The spilled oil quickly formed tar balls that moved below the water surface (i.e., were overwashed by waves) and could not be tracked visually. Two weeks later, oil unexpectedly came ashore along the coast of Vancouver Island, Canada, 175 kilometers north of the release site, contaminating 150 kilometers of shoreline (NOAA, 1992). The oil had a significant effect on the large number of marine birds wintering in the area. Of the 10,300 birds collected, about 9,300 were either already dead or died in treatment centers. Many more were believed to have died but were never collected.

troughs of offshore bars. These rollers picked up more sand and shell fragments as they moved, making them heavier.

Experts have long been concerned that oil spilled in turbulent waters with heavy loads of suspended silt and clay (i.e., glacial meltwater, such as upper Cook Inlet and the Yukon River) would mix with the sediments and sink (Kirstein et al., 1985). Laboratory studies have shown that oil mixed with water with heavy suspended sediment loads does adhere to the sediments, with concentrations up to 0.1 gram of oil per gram of solid (McCourt and Shier, 1998). However, this process is likely to result in the deposition of oiled sediments rather than the transport of bulk oil to the bottom.

During the Tampa Bay and *Morris J. Berman* spills (Box 2-2), response teams observed that floating oil sank by mixing with sand in nearshore waters, without coming into contact with intertidal sediments on the shoreline (Figure 2-3c). If a floating slick of heavy oil drifts into shallow water along an exposed shoreline, it is more likely to be mixed into the water column by wave turbulence. If the bottom is sandy, the sand may be suspended in the water column by waves and could mix with the oil. The suspended sand concentrations in breaking waves is commonly 300 to 500 mg/L and can easily reach 5,000 mg/L (Kana, 1979), compared to typical concentrations of fine-grained suspended sediment of 20 mg/L in estuaries and nearshore waters. Because the specific gravity of quartz is 2.65, it only takes 2 to 3 percent sand by weight mixed into oil to make it heavier than seawater. Again, high viscosity is an important factor, because viscous oils tend to form large tar balls (rather than small droplets) that pick up sand. The oil/sand mixture can be carried by long shore currents and deposited in relatively sheltered areas where it can form extensive, thick layers of oil/sand on the bottom.

BOX 2-2
The *Morris J. Berman* Spill

On January 7, 1994, the *Morris J. Berman* barge ran aground just offshore San Juan, Puerto Rico, releasing about 18,000 barrels of heavy fuel oil (API gravity of 9.5). Although much of the oil floated, response teams reported finding oil on the bottom within the first 24 hours, and eventually mats of submerged oil were found in both offshore areas and on the landward side of nearshore reefs. Most of the sunken mats were within 1 or 2 kilometers of the vessel, although one site was 110 kilometers from the release site. The oil adhered to rocky surfaces and coated seagrass beds (Burns et al., 1995). It was later determined that most of the oil on the bottom had sunk without coming into contact with the shore (Michel et al., 1995). The oil contained a few percent sand and could readily refloat in seawater and recontaminate the adjacent shoreline once it was separated from the sand. Three different methods were used to remove the oil: diver-directed vacuuming of the more liquid oil; manual pickup by divers of the more viscous patches; and dredging of large deposits in a small bay (Burns et al., 1995; Ploen, 1995).

Oil Heavier than Water, Low Currents

If the density of the oil is higher than the density of the receiving water, some of the oil can form a sheen, but the majority does not float. As the oil mixes into the water column, it forms small droplets, ranging in size from approximately 0.5 microns to several millimeters. If the water column is strongly stratified, some of the oil droplets may accumulate on the pycnocline, provided that they are less than the underlying water. If current speeds are low, oil that is more dense than the water sinks and accumulates on the bottom (Figure 2-3d). Direct sinking in low-flow areas was observed after the *Sansinena* oil spill (see Box 2-3) while it was docked at a pier (Hutchison and Simonsen, 1979), and the *Mobiloil* spill (in the lee of the grounded vessel) (Kennedy and Baca, 1984).

Suspended oil can sink when the oil is transported into low-flow areas similar to the way fine-grained sediments are deposited in estuaries during slack periods of the tide. However, oil droplets can be readily remobilized by tidal currents, so long-term accumulation is likely only in areas where wave-generated, tidal, or riverine currents have little effect. Examples of such areas include abandoned channels, dredged channels or pits, depressions adjacent to piers caused by "propeller wash" of anchoring vessels, dead-end canals, and the lee side of natural and man-made structures. If the oil does accumulate on the bottom, the oil droplets recoalesce into pools of liquid oil that can be tens of centimeters thick. Evaporation and photo-oxidation of sunken oil are much slower than for floating oil slicks, and the oil tends to remain as a liquid on the bottom. Dissolution from thick mats is slow (Lee et al., 1989). Observations of spills have shown that this

> **BOX 2-3**
> **The *Sansinena* Spill**
>
> On December 17, 1976, the tanker SS *Sansinena* exploded while loading fuel in Los Angeles harbor, releasing more than 33,000 barrels of bunker fuel oil (API gravity 7.9 to 8.8). Approximately 200 barrels floated, but the majority of the oil sank. Divers reported large pools of oil up to three meters deep on the harbor bottom, where the oil had settled into depressions (Hutchison and Simonsen, 1979). Initial recovery was by diver-directed vacuum removal and separation in tanks mounted on a barge, but this method was abandoned because of the great difficulty of moving the suction head along the uneven bottom. Next, diver-guided hydraulic pumps were used on thick accumulations close to the pier. Specially designed pumping units consisting of a prime mover and hydraulic pumps on a barge were then used to collect oil from outer depressions. Nearly 16,000 barrels were recovered during the initial recovery operations. Eventually, a suction head and pump device was designed on site for recovery of the large quantities of oil still remaining on the bottom. This device had to be operated according to directions from a diver because some of the oil pools had become silted over and even had marine life living in the silt, making the oil difficult to locate. During the next 90 days, 10,300 barrels were recovered from the harbor bottom. Over a 16-month period, 33,000 barrels, nearly all of the spilled volume, were recovered.

type of oil does not initially adhere to or mix with large amounts of fine-grained sediments under water.

Oil Heavier than Water, High Currents

If currents are greater than about 0.1 m/s, oil droplets stay suspended in the water column and disperse (Figure 2-3e). In rivers and most nearshore coastal settings, the oil is not likely to accumulate on the bottom because the currents are strong enough to keep it suspended in the water column. For example, little or no oil accumulation on the bottom was observed after heavy-oil spills in the Columbia River (Kennedy and Baca, 1984), the Mississippi River near Vicksburg (Weems et al., 1997) and in Puget Sound (Yaroch and Reiter, 1989). However, even in strong currents, heavy oils can accumulate in sheltered areas. For example, after about 4,760 barrels of slurry oil were spilled into the Mississippi River, nearly 50 percent of the oil was recovered from the bottom, but only from the lee created when the leaking barges were pushed at a 45-degree angle against the river bank (Weems et al., 1997). No other significant amounts of oil were found in extensive surveys. The oil was not expected to adhere initially to debris or vegetation as it mixed into the water column because fresh oil generally does not stick to water-wet surfaces.

Spills of Emulsified Fuels

Emulsified fuels (anthropogenic fuels manufactured by mixing water and surfactants with liquid oils or solid hydrocarbon products) behave very differently. Because only one small accidental spill of emulsified fuel has been reported (Sommerville et al., 1997), our understanding of the behavior of these oils is based mostly on research conducted specifically with Orimulsion™, an emulsified fuel manufactured from bitumen produced in Venezuela. Laboratory and field experiments on emulsified oils have been conducted in Canada (Jokuty et al., 1995), the United States (Deis et al., 1997; Ostazeski et al., 1997), Venezuela, and Europe (Sommerville et al., 1997). In freshwater, the surfactant in emulsified fuels will maintain its effectiveness over longer periods of time, preventing recoalescence of the bitumen particles. In low-flow conditions (Figure 2-4a), the spilled oil will settle to the bottom of the water column. In these quiescent conditions, the oil has little potential for mixing with sediment, except in the long term by bioturbation.

In freshwater with currents, the predispersed bitumen particles will slowly descend to the bottom down current (Figure 2-4b), and the surfactant will remain effective for a limited period of time, preventing recoalescence of the particles. The eventual fate of the bitumen particles is uncertain, particularly in terms of interaction with fine-grained sediments. Because the bitumen particles are highly adhesive, it is likely that they will adhere to suspended sediments and eventually be deposited in low-flow zones.

In saltwater, the emulsified oils will initially form clouds of dispersed particles in the upper 1 or 2 meters of the water column (Figure 2-4c). Laboratory and field tests have shown that surfactants quickly lose their effectiveness in saltwater. In areas with high bitumen concentrations, the particles can recoalesce and rise to the surface, forming tarry slicks. In wave-tank experiments, the tar coated the glass sides of the wave tank (Jokuty et al., 1995). However, in open water, the particles would disperse. Therefore, options for containing and recovering spilled emulsified oils quickly decrease over time.

Refloating Mechanisms

Sunken oil can refloat, creating significant problems for spill-response teams and a chronic source of exposure. In the *Morris J. Berman* spill, months after the spill large quantities of liquid oil refloated, recontaminating beaches and exposing resources in the water-column and on the surface to oil after the bulk of the floating oil had been recovered. There are three mechanisms for refloating oil: (1) still-buoyant oil can separate from the sand; (2) wave-generated currents can

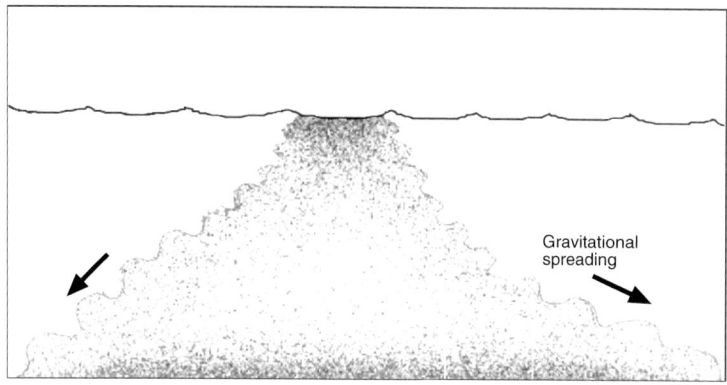

FIGURE 2-4a Emulsified oil in freshwater; low currents; oil sinks.

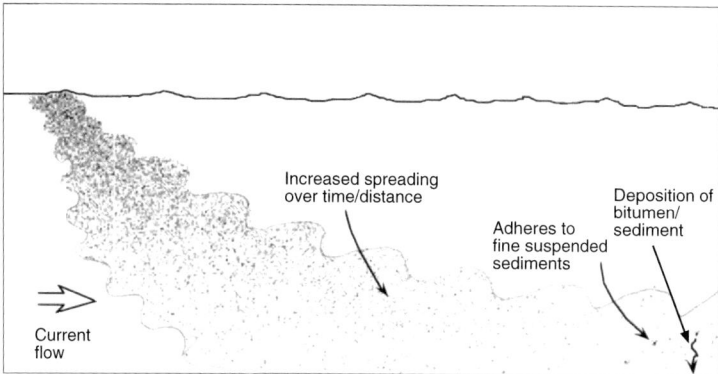

FIGURE 2-4b Emulsified oil in freshwater; high currents; oil disperses and eventually sinks.

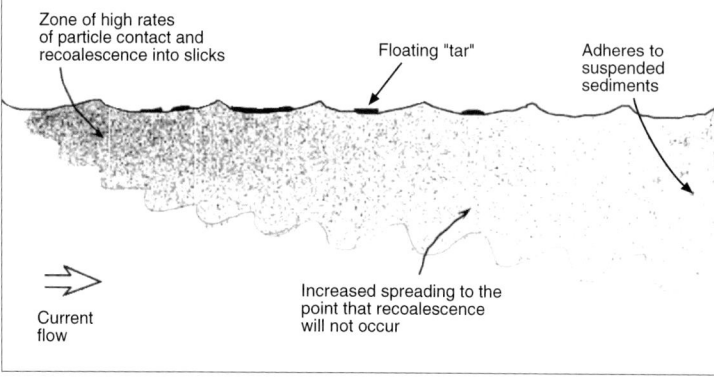

FIGURE 2-4c Emulsified oil in saltwater; high currents; oil initially disperses then coalesces into tarry slicks.

loosen and resuspend pieces of buoyant oil from the bottom; and (3) increases in water and/or oil temperature can make the oil less viscous and/or more buoyant.

Spill-response teams often assume that oil refloats because of a short-term change in temperature (e.g., in the afternoon when the water warms up). However, it is unlikely that short-term changes in temperature can cause oil-sand mixtures to sink in one situation and float in another because differential oil and water expansion coefficients are very small. Hence, oil on the bottom and the overlying water expand at about the same rate, and relative changes in density are small.

A more likely mechanism for refloating sunken oil is the physical separation of the sediment from oil (Michel and Galt, 1995). If one dumps sand into a can of motor oil, the sand falls through the oil by gravity and forms a layer of sand on the bottom. Settling rates through more viscous oils would be longer but could be increased in the field by wave motion and other physical processes. In the *Morris J. Berman* spill, buoyant oil droplets were observed breaking off layers of oil on the bottom, somewhat like droplets being released from the heated mass in lava lamps. These droplets are believed to have been formed as the still-buoyant oil became less viscous during the daytime heating of the water, allowing the oil to separate from the sand and droplets to break away from the submerged oil by wave action generated by the land-sea breezes.

POTENTIAL EFFECTS OF NONFLOATING-OIL SPILLS

When a floating oil is spilled, response teams typically have to recover oil slicks, clean up oil stranded on the shoreline, and recover and treat animals along the shoreline and in the water. Their focus is on the water surface and shoreline, the so-called "bath-tub ring." Life forms in the water column and benthic habitats are usually considered to be at less risk of exposure and injury from floating oil slicks than from nonfloating oils. Table 2-1 compares the predicted impacts of nonfloating-oil spills and floating-oil spills on shoreline and benthic habitats, major assemblages of fish and wildlife, and human-use resources. Spills of nonfloating oils are expected to have less impact on shoreline habitats because smaller amounts of oil are likely to be stranded and cleanup activities are likely to be less disruptive (Scholz et al., 1994). Any oil that is stranded, however, is likely to be very persistent because of the slow natural removal rates for heavy, adhesive oils. Nonfloating oils are less likely to penetrate porous sediments or wetland vegetation because of their high viscosities and adhesiveness (Harper et al., 1995).

Impacts on water-surface resources are also expected to be lower from spills of nonfloating oils because of the significant reduction in the amount of oil on the water surface. If the oil refloats, it could be a chronic source of exposure to both water-surface and shoreline resources, but the risk is likely to be limited to areas adjacent to sunken oil deposits (NOAA, 1995).

All water-column and benthic habitats are at increased risks from spills of

TABLE 2-1 Relative Changes in the Resources at Risk from Spills of Nonfloating Oils Compared to Floating Oils

Resource at Risk	Risks from Spills of Nonfloating Oils Compared to Spills of Floating Oils
Rocky Shores (−)	Less oil is likely to be stranded, but oil that is stranded is usually stickier and thicker.
Beaches (−)	Viscous oils are less likely to penetrate porous sediments. Oil is often stranded as tar balls, which are easy to clean up on sand beaches. Chronic recontamination is possible for months.
Wetlands and Tidal Flats (−)	Less oil coats vegetation. Because the oil does not refloat with the rising tide, any oil stranded on the lower intertidal zone will remain, thus increasing risk to biota. Cleanup of oil from these environments is very difficult, and natural recovery takes longer.
Water Surface (−)	Less oil remains on the water surface. Oil tends to form fields of tar balls. Potential for chronic impacts from refloated oil over time is high.
Water Column (+)	Oil can increase exposure as it mixes in the water column. Risks increase if oil refloats after separation from sediments. When submerged, slow weathering of the more toxic components can be a chronic source of risk.
Benthic Habitats (++)	Risks are significantly increased for areas where heavy oils accumulate on the bottom. Slow weathering rates further increase the risk of chronic exposures. Smothering and coating can be heavy. Bioavailability varies with oil and spill conditions.
Birds (−)	Less oil remains on the water surface, so direct and acute impacts are lower. There is a high probability of chronic impacts from exposure to refloated oil and restranded tar balls on shores after storms.
Fish (+)	Risks are increased to all fish, especially benthic or territorial fish, in areas where oil has accumulated on the bottom.
Shellfish (++)	Risks are increased to all shellfish, especially species that spend most of their time on the sediment surface (e.g., mussels, lobsters, crabs). Risk of chronic exposure from bulk oil, as well as the slow release of water-soluble PAHs (polynuclear aromatic hydrocarbons), is high.
Marine Mammals (−)	Less oil remains on the water surface, and the potential for contamination of marine mammals on shore is lower. Oil in the water column is not likely to have an impact on highly mobile species. Benthic feeders (such as manatees) could be exposed from accumulations on the bottom, which would weather slowly.
Sea Turtles (−)	Less oil remains on the water surface, and less oil is stranded on nesting beaches.
Water Intakes (++)	Oil mixed into the water column would pose serious risks to water treatment facilities. Closures are likely to be longer.

Note: (−) indicates a reduction in risk. (+) indicates an increase in risk. Actual risks for a specific spill will be a function of the composition and properties of the spilled oil and environmental conditions at the spill site.

nonfloating oils (Scholz et al., 1994). Oils that quickly sink or are suspended in the water column have greater impacts on organisms in the water column because more of the water-soluble fraction of the oil dissolves rather than evaporates. Oil on the surface is primarily weathered by evaporation to the atmosphere and, to a lesser degree, to the water column by dissolution. Oils suspended in the water column or deposited on the bottom are less likely to evaporate but more likely to dissolve, although the water-soluble fraction of heavy oils is usually very low. Consequently, the water column can have higher concentrations of toxic fractions from nonfloating oils than from floating oils. Dissolution tends to be a slower process than evaporation (Lee et al., 1989, p.37), thus increasing potential exposure times. In the *Morris J. Berman* spill in Puerto Rico, divers observed dead fish, living fish with lesions and tumors, and many lethargic territorial fish in nearshore waters adjacent to the spill site (Vincente, 1994). Mobile species may be able to move to uncontaminated areas, thus reducing their exposure.

Nonfloating oils are often high in polynuclear aromatic hydrocarbons (PAHs), which are the primary source of both acute and chronic toxicity to aquatic organisms. Naphthalene compounds (two-ringed aromatics) have been shown to be more toxic than lightweight aromatics, such as benzene and toluene (Anderson et al., 1987). In terms of the water-soluble fraction, bunker C is as toxic as diesel oil (Markarian et al., 1993). Thus, even though heavy residual oils are not usually considered to be acutely toxic to fish (NOAA and API, 1995), oils that are mixed into the water column without weathering by evaporation on the water surface first may have a higher fraction that dissolves and, therefore, may be more acutely toxic to organisms in the water column.

3

Technologies and Techniques

In this section, the current technologies and techniques for locating, tracking, containing, and recovering spills of nonfloating oils are summarized. The presentation is divided into subsections on spill modeling and information systems, spill tracking and mapping, and oil containment and recovery. The summary focuses on the current state of practice and identifies systems that have been applied or proposed for application to submerged oil. Summaries of the use of these techniques in selected spills in which substantial quantities of oil were submerged or deposited on the seabed can be found in Michel and Galt (1995) and Michel et al. (1995). An annotated bibliography of the literature can be found in NOAA (1997).

MODELING AND INFORMATION SYSTEMS

The following discussion begins with a brief overview of the state of the art in spill modeling and information systems (Box 3-1). This is followed by the extension of spill models to include the subsurface transport and deposition of dispersed oil and a history of the use of these models to "hindcast" (analyze a past event) several large accidental spills in which subsurface transport was important. The use of models to forecast and hindcast spills involving substantial amounts of submerged oil is then summarized.

Recent comprehensive reviews of the state of the art in spill modeling (Spaulding, 1995; ASCE, 1996) show that the models have evolved quite rapidly taking advantage of the availability of low-cost, high-powered workstations and personal computers with full color graphics, extensive storage, and communications systems. A simultaneous evolution in the software has enabled a clear

BOX 3-1
Oil-Spill Model

The core of an oil-spill model is a series of algorithms that represent the processes controlling the transport and fate of oil released into the environment. The transport portion of the models describes the physical movement of oil by winds, currents, waves, and associated turbulence. The fate of the oil is normally represented in terms of spreading, evaporation, dispersion or entrainment, dissolution, emulsification, biodegradation, sinking or sedimentation, photo-oxidation, and oil-shoreline and oil-ice interactions. These processes are typically formulated individually with links to other processes or environmental data as necessary to describe the oil's fate. The algorithms may be altered or changed entirely depending on the environment in which the oil is spilled or transported.

Input to oil-spill models normally includes a description of the study area, the oil-spill scenario (spill location, release rate and schedule, and oil type), and environmental conditions. The study area is normally described using a map of the region of principal interest. The environmental forcing data typically consist of estimates of the temporally and spatially varying wind and current fields for the forecast period (typically a few days for spill-response support) and an estimate of the mean water temperature. These environmental data fields may be provided by supporting hydrodynamic and meteorological models for the study area or from observations. The model output typically includes animations of the movement of the surface oil and the oil mass balance by major environmental compartments (surface, water column, onshore, evaporated, seabed, biodegraded), the oil thickness and areal extent, and the oil properties (viscosity, water content) versus time.

---

separation to be made between the model software and supporting environmental data (Spaulding and Chen, 1994). With model/data separation, the models can be rapidly applied to new locations (Anderson et al., 1993). Many models have been linked with geographic information systems (GISs) or have limited GIS functions embedded in the model systems (Galagan et al., 1992). With the incorporation of the GIS and other data management tools, users can input, organize, manipulate, archive, and display georeferenced information relevant to spill modeling. With the extension of spill models to include supporting data management tools, spill information systems have been developed that can provide valuable data to support spill responses and planning.

In most cases, models have been tested and validated by application to selected, usually large, accidental spills or experimental field trials. These events are selected based on the availability and quality of data. Hindcasts of the largest, most recent spills (*Exxon Valdez*, the Gulf War spill, *Braer, North Cape*) have been used by several researchers to demonstrate the predictive performance of their models.

Basic spill models have been extended to include biological and, in some cases, economic models for estimating the impact and damages of spills (e.g., French et al., 1994). These models are now being incorporated into comprehensive, on-scene, command-and-control systems (Anderson et al., 1998). Strategies for using models to prepare a trajectory analysis have been developed by Galt (1994, 1995). The National Oceanic and Atmospheric Administration (NOAA) has also developed digital distribution standards for data on trajectories (Galt et al., 1996).

Most of the spill models developed to date focus on the transport and fate of surface oil slicks. These models typically predict the mass of oil removed from the sea surface by evaporation, by dispersion or entrainment into the water column, and by sinking and sedimentation but do not explicitly track the dispersed oil. This approach has been taken because most spills involve oils that float throughout most of the short-term spill response. Selected models have the capability of predicting the three-dimensional evolution of oil, including entrainment, subsurface transport, sedimentation, and refloating of spilled oil (e.g., Spaulding et al., 1994; Elliot, 1991; Johansen, 1985; French et al., 1994). The majority of these models employ a particle-based, random-walk technique to predict the evolution of subsurface oil (Kolluru et al., 1994) although other alternatives have also been investigated (Spaulding et al., 1992). In these models, the influence of oil sediment interaction (Kirstein et al., 1985) and the buoyancy of dispersed oil droplets are explicitly accounted for.

The use of the three-dimensional models to forecast and hindcast spills has been limited. Most simulations have been restricted to buoyant oils that have been dispersed in the water column by strong winds or wave forcing. Although these oils are not a direct analog for nonfloating oils, they are instructive in illustrating the ability to predict the transport and fate of oil dispersed in the water column. For example, both Proctor et al. (1994) and Spaulding et al. (1994) performed hindcasts of the *Braer* spill. Both models correctly predicted the general subsurface transport of the highly dispersible, Gulfaks crude oil that was spilled. The predicted location of the subsurface oil was consistent with the pattern of sedimented oil found on the seabed. Neither hindcast included oil-sediment interaction, however, and no predictions were made of the deposition of sedimented oil.

A review of the literature on oil beneath the water surface and Group V oils by NOAA (1997) shows that spill models have generally not been used to forecast or hindcast spills of heavy oils. This is consistent with the summaries of spills of heavy oils presented in Michel and Galt (1995) and Michel et al. (1995). The absence of model applications to forecast or hindcast these events can be attributed to several factors. First, spills of heavy oils are generally less frequent, and the volume of oil spilled tends to be less than in spills of floating oils. Second, requirements for current data (either from observations or hydrodynamic model predictions), which are difficult to obtain for surface spills, are increased substantially when the subsurface transport of oil is involved. The subsurface current structure is of limited importance when the flows are principally tidal and water depths are shallow, but they become particularly important when stratification and multilayer flows are present. Finally, Michel and Galt (1995) have shown that substantial subsurface transport and deposition often occur as the result of the interaction of buoyant oil with sand. The sinking and subsequent deposition of oil caused by changes in the oil's density due to weathering (evaporative losses) or burning are rare (Lee et al., 1989, 1992).

Most spill models are focused on predicting the transport and fate of oil at sea and do not include oil-sediment interactions or oil-shoreline interactions. Given the lack of data and the lack of a clear understanding of the controlling processes, those that do are necessarily rudimentary (ASA, 1997; Reed et al., 1989). Incorporating oil-sediment interactions into spill models will require estimates of the suspended sediment concentrations as input (Kirstein et al., 1985). These estimates are normally based on observations or model predictions, and the data are rarely available during spill events. Incorporating oil-shoreline interactions will require extensive data on the nearshore environment, including geomorphology and wave and current fields. Once again these data are generally not available for most spills, particularly during the emergency response phase.

Given this situation, two strategies might be tried to use existing spill models to assist in the response to spills where subsurface transport processes and sinking and sedimentation might be important. First, the spill model could be used to explore the impact of various assumptions about the subsurface transport of the oil and the interaction of oil and sediment. For example, it could be assumed that a portion of the oil will be removed or leave the surface as it becomes neutrally bouyant or sinks at a specified rate due to oil-sediment interaction. Model predictions could then be made to estimate the path and a general sense of the area and volume that would be impacted by the subsurface oil. The information could be used to establish field sampling programs. Data collected from the field on the current structure and sediment concentrations could then be used to refine the predictions and narrow the scope of the uncertainty.

A second approach would be to place the spill model in real-time operation for the principal areas of concern. Supporting three-dimensional hydrodynamic and sediment-transport models for nearshore and offshore areas would provide currents and suspended-sediment fields for inputs to the spill model. The models, which would have been validated with field observations, would be able to assimilate real-time data from monitoring systems to maximize their predictive performance. This approach would only be viable for areas where the probability of spills is high enough to warrant the investment in the development, application, and maintenance of such a system.

TRACKING AND MAPPING TECHNIQUES

Techniques for tracking and mapping the location of oil throughout a spill and subsequent cleanup are critical to the effective containment and recovery of oil in the water column or deposited on the seabed. A brief summary of current methods for tracking and mapping subsurface oil follows. The review is based primarily on summaries in Castle et al. (1995) and Michel et al. (1995). Additional information is available in Smedley and Belore (1991) and Brown et al. (1997). As a practical guide to determining which tracking and mapping options are most appropriate, Figure 3-1 provides a typical decision tree based on oil density and water depth. The first branching is based on assessing the density of oil relative to the density of the receiving water and includes two branches, one if the oil is neutrally buoyant and one if the oil is negatively buoyant in receiving water. The second branching depends on the water depth. Final selection of the tracking method is dependent on local conditions, the availability of equipment and personnel, and weather conditions.

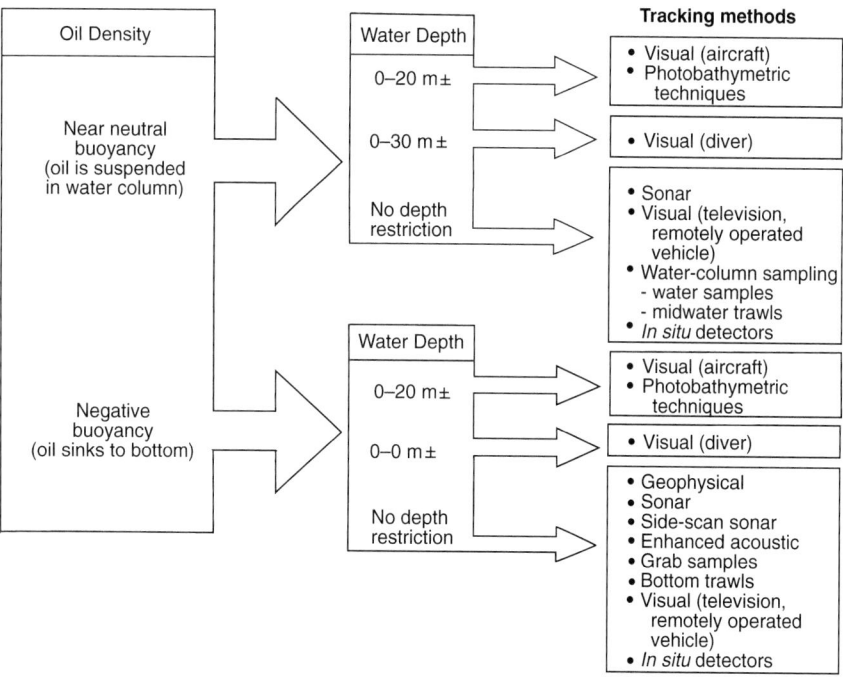

FIGURE 3-1 Decision tree based on oil density and water depth. Source: Castle et al., 1995.

Visual Observations

Visual observations (by aircraft, ship, diver, or camera/television) have been the principal methods of locating and tracking submerged oil. Airborne photography and visual-based systems, which are widely available and can rapidly survey large areas, are widely used to locate submerged oil. The performance of these systems is limited by water clarity and depth, the quantity of oil, and the characteristics of bottom sediment. Given the possibility of misidentifying natural materials (seaweed, seagrass beds) as oil, *in situ* observations are always required to validate airborne assessments. Direct observations can also be performed by divers within safe depth restrictions and visibility limits. Observations by underwater cameras, either operated by divers or deployed from ships, can also be used to locate submerged oil. These visual methods must generally be confirmed by sampling and have relatively limited coverage. As an extension of visual methods, photobathymetric techniques, such as multispectral photography, may be useful for mapping oil on the seabed in shallow water (Benggio, 1994b). Once again, field confirmation and calibration are required.

Remote Sensing Techniques

Standard, side-looking, airborne radar, synthetic-aperture radars, and infared/ultraviolet line scanners are generally unable to map subsurface oil because they cannot penetrate the water surface (Fingas and Brown, 1996). The methods are also hindered by the weather and visibility. Laser fluorosensor techniques have been developed and shown to be able to detect oil in the water column for the purposes of oil exploration (Dick and Fingas, 1992; Dick et al., 1992). Little evidence exists that this technique has been used in responding to spills of nonfloating oils, however (Brown et al., 1997). Recent laboratory experiments by Brown (1998) have demonstrated a laser airborne fluorosensor that can detect the presence of dispersed bitumen in the water. No field tests or practical uses of the system have been made to date.

Geophysical/Acoustic Techniques

These technologies include of a variety of acoustic-based techniques for locating and mapping submerged oil (Chivers et al., 1990). These techniques rely on acoustic sounding principles, specifically the differential density and sound speeds of water compared to those of oil or oil-sediment mixtures and the scattering of sound waves from particulate material in the water column. Oil in the water column can be qualitatively mapped by commercial fish-finding and echo sounders or by precision survey equipment. Oil on the seabed and associated bottom features can be mapped by side-scan sonar systems. The output of these systems can be enhanced for mapping the texture and composition of the bottom.

One such system was reportedly used to map the submerged oil from the *Morris J. Berman* and *Haven* spills (Marine Microsystems, 1992).

Side-scan sonar mapping systems are normally interfaced with the global positioning system (GPS) and hydrographic mapping software to generate maps of seafloor features. These systems can provide relatively rapid coverage but are most useful when they are used to direct the surveys for areas of natural collection that have already been identified. These specialized systems may be unable to distinguish between oiled sediments and underlying sediments because of their acoustic similarity. Therefore, sampling or *in situ* observations are necessary to confirm the maps.

Water-Column and Bottom Sampling

Direct sampling of the water column or seabed may be used to locate and map the movement of oil. Sampling can be done by a vessel, a remote vehicle, or a diver (in shallow water). Sampling generally becomes more difficult and time consuming as the water depth, current speed, and wave height increase. A variety of sampling techniques are available, including grab sampling of water or sediments with subsequent visual or chemical analysis, sorbent materials deployed on weighted lines or in traps (Benggio, 1994a), and core sampling of the seabed sediments. Sampling is typically limited in scope and may not provide representative observations of the impact area. Water-column and bottom trawls may be useful for selected spills because they can cover larger areas. The effectiveness of sampling methods is strongly dependent on the composition of the oil and oiled sediment and environmental factors, such as current speed, water depth, and substrate type.

In Situ *Detectors*

In situ and towed fluorometric detection are widely available and routinely used to detect and map petroleum leaks and spills (Turner Designs, 1999). These systems may be mounted on buoys, boats, or remotely operated vehicles. When mounted on boats and coordinated with GPS, they can provide maps of the subsurface oil concentration field. They are restricted to making oil concentration measurements in the water column (Brown et al., 1997) and have a detection range from parts per billion to parts per million, depending on environmental conditions and oil type. Given the three-dimensional nature of submerged oil plumes, mapping of subsurface oil requires an extensive effort. Towed systems might also be used to monitor conditions at one location, such as in a river, to determine whether oil has reached that location and is being transported downstream. These systems have historically been used to assess the effectiveness of dispersants in field trials and planned spill events. They have not been routinely used for actual spills in the United States but are used in Canada and the United Kingdom to assess the potential for tainting fish from subsurface oils.

Summary

The appropriate method for tracking and mapping a particular spill depends on whether the oil is suspended in the water column or deposited on the seabed and on the water depth and clarity. In general, visual and photobathymetric techniques are restricted to water depths of 20 meters or less and are suitable for both suspended and deposited oil. Diver-based visual observations can only be used in low-current and small wave areas. Acoustic techniques, television observations, water-column and bottom sampling, *in situ* detectors, and nets and trawls typically have no depth restrictions except that the water must be deep enough for the instrument to be deployed and operated safely. They become more difficult to operate, however, as the current speed and wave height increase. Measurements near the seabed become more challenging as the topographic relief of the bottom increases and the bottom surface becomes rougher. Tables 3-1 and 3-2 provide a summary of the uses and limitations of various tracking and mapping methods.

CONTAINMENT AND RECOVERY METHODS

The following descriptions summarize the current state of practice for containing and recovering heavy oils. The summary is based principally on work by Michel et al. (1995), Castle et al. (1995), and Benggio (1994c). Additional information is available in Bonham (1989), and Moller (1992). A useful summary of the containment and recovery of sinking hazardous chemicals is presented in Boyer et al. (1987). Brown et al. (1997) provide a useful summary of the practical aspects of containing and recovering spills of "sunken and submerged oils" and also summarize the methods used in successful responses to spills. Supporting data on these successful responses can be found in NOAA (1997).

Protocols for determining which methods to use for a given spill situation have been proposed by Castle et al. (1995). The approach is based on a decision tree structure, with the principal branching being determined by the buoyancy of the oil, the depth of the water column, and whether the oil is pumpable or not. Figures 3-2 and 3-3 show decision trees for the containment and recovery of sunken oil, respectively. Criteria for each branch are also provided. The form of the decision tree is similar to the one for tracking and mapping (see Figure 3-1).

Containment

Oil that is spilled and transported subsurface either remains suspended in the water column or is deposited on the seabed, usually after interaction with suspended sediments or sand. Different strategies for containing these oils can be used depending on the location of the oil. Typical response strategies are described below. Few of these techniques have been used and their performance has not been documented during spill events.

Oil in the Water Column

Silt Curtains. The containment of oil suspended in the water column is generally possible only in areas with weak currents (less than 10 cm/sec) and small waves (less than 0.25 m). Silt curtains, which are normally used to control the transport of suspended sediment during dredging operations, are typically restricted to water depths of 3 to 6 meters and are deployed so that the bottom of the curtain does not extend to the seabed. They have not been used in actual spill events.

Nets and Trawls. Midwater trawls and nets may be used for containing selected oil types in certain conditions. The performance of these systems depends on the viscosity of the oil and being able to locate and concentrate the oil. Delvigne (1987) has suggested that nets can successfully contain oil if the currents are low (less than 10 cm/sec) and the viscosity of the oil is high. Nets can be towed, moored, or mounted on moving floats. This method is sometimes used to protect fixed structures (water intake systems) or resources at risk. The effectiveness of trawls and nets declines rapidly as current speeds increase or as nets become clogged. During the *Presidente Rivera* spill in the Delaware River, fish nets were able to recover eight tons of oil before they became fouled (NOAA, 1992).

Pneumatic Barriers and Booms. Pneumatic barriers involve injecting air at the seabed and forming a bubble plume that rises to the surface. Pneumatic barriers have been considered for protecting seawater intakes against oil dispersed in the water column, but little data are available for assessing their performance. Standard oil booms (deep draft) have been considered for containing subsurface oil. In fact, booms have been suggested as the preferred option for responding to spills of bitumen-surfactant-water mixtures and have undergone limited testing at sea (Deis et al., 1997; Sommerville et al., 1997). Booms can be used only when the oil remains in the upper water column, the currents are low (less than 0.20 m/sec), and the waves are small (less than 0.25 m).

Oil on the Seabed

Seabed Depressions. Oil deposited on the seabed can be moved by ambient currents and waves. Sedimented oil tends to collect in natural or man-made depressions on the bottom, including natural and dredged channels, wave-generated troughs offshore of sandy beaches, and natural depressions. Dredging to create depressions for oil collection is not practical as part of a spill response except for very large spills or spills that have very substantial benthic impacts.

TABLE 3-1 Options for Tracking Oil Suspended in the Water Column

	Visual Observations	**Water Sampling**
Description	Trained observers in aircraft or on vessels look for visual evidence of suspended oil; includes use of cameras.	Visual inspection or chemical analysis of grab water samples or a flow-through system with a fluorometer.
Availability of Equipment	Uses readily available equipment.	Uses readily available equipment and supplies.
Logistical Requirements	Low/aircraft and vessels are readily available during spill response.	May require boat, sampling equipment, pumps, GPS for station location, portable oil analyzer.
Coverage Rate	High for aircraft; moderate for vessels.	Very low coverage rate; collecting discrete water samples at multiple depths for testing is very slow.
Data Turnaround	Quick turnaround.	Quick turnaround for visual analysis; chemical results would have to be available in minutes to be effective.
Probability of False Positives	High probability, due to poor water visibility, cloud shadows, seagrass beds, irregular bathymetry, mixing of different waterbodies.	Low probability; field personnel would have to know how to operate all equipment.
Operational Limitations	Requires good water visibility and light conditions; poor weather may restrict flights; limited to daylight hours.	Realistic only for water depths <30 ft; sea conditions may restrict vessel operations.
Pros	Can cover large areas quickly using standard resources available at spills.	Can be used at points of concern, such as water intakes.
Cons	Only effective in areas with very low water turbidity.	Too slow to be effective in dynamic settings or over large areas.

TECHNOLOGIES AND TECHNIQUES 43

Fish Net Trawls	**Sorbent Fences**	**Airborne Imaging LIDAR**
Fish nets or trawling gear are towed for set distances then inspected for presence of oil; or nets can be set at fixed points and regularly inspected.	Sorbents are attached to something like a chain link fence which is submerged into the water then pulled for inspection; or it could be set at a fixed point for regular inspection	Pulsed laser and video recording system compares back-reflectance from below the water surface for areas of suspended oil versus clean water. Detection depth varies (nominally 45 ft). Operable 24 hours/day
Readily available in commercial fishing areas.	Uses readily available equipment and supplies	Uses very specialized equipment of limited availability
Moderate; requires boat and operators to tow the nets; may require multiple vessels to cover large areas; may require many replacement nets as they become oiled.	Low; can be deployed from small boats or carried to small streams for deployment	Moderate; equipment must be modified for mounting on local aircraft; requires skilled operators
Low coverage; nets have a small sweep area and must be pulled frequently for inspection.	Low; they have a small sweep area and they have to be pulled frequently for inspection	High; flown on aircraft with 200 ft swath
Quick turnaround.	Quick	Moderate; data recorded on video
Low probability; oil staining should be readily differentiated from other fouling materials.	Low; sorbents are designed to pick up oil, so they would be less likely to be stained by other materials	High; system images all submerged features, have to learn to identify patterns for different features, thus requires extensive ground truthing
Obstructions in the water can hang up nets; restricted to relatively shallow depths; sea conditions may restrict vessel operations.	Difficult to deploy and retrieve in strong currents; sea conditions may restrict vessel operations	Weather may restrict flights; minimum detectable size of oil particle is not known, but other individual features detected are usually feet in size or schools of small fish
Can sweep various depths or very close to the bottom.	Uses material available anywhere	Can cover large areas quickly using standard resources available at spills; permanent record of image that is geo-referenced
Very slow; nets can fail from excess accumulation of debris.	Very slow; very limited sampling area	Not proven for detecting suspended oil droplets; very limited availability

TABLE 3-2 Options for Mapping Oil Deposited on the Seabed

	Visual Observations	Bottom Sampling from the Surface	Underwater Surveys by Divers
Description	Trained observers in aircraft or on vessels look for visual evidence of oil on the bottom; includes underwater cameras.	A sampling device (corer, grab sampler, sorbents attached to weights) is deployed to collect samples from the bottom for visual inspection.	Divers (trained in diving in contaminated water) survey the sea floor either visually or with video cameras.
Availability of Equipment	Uses readily available equipment.	Uses readily available equipment and supplies	Underwater video cameras are readily available, but divers and diving gear for contaminated water operations may not be available locally.
Logistical Needs	Aircraft and vessels are readily available during spill response.	Requires boat, sampling equipment, GPS for station location.	Depend on the level of diver protection required.
Coverage Rate	High for aircraft; low for vessels.	Very low coverage; collecting discrete bottom samples is very slow; devices sample only a very small area.	Low coverage, because of slow swimming rates, limited diving time, poor water quality.
Data Turnaround	Quick turnaround.	Quick turnaround because visual analysis is used.	Quick turnaround.
Probability of False Positives	High, due to poor water clarity, cloud shadows, seagrass beds, irregular bathymetry.	Low probability, except in areas with high background oil contamination.	Low probability because divers can verify potential oil deposits.
Operational Limitations	Requires good water clarity and light conditions; weather may restrict flights; can be used only during daylight hours.	Sea conditions may restrict vessel operations.	Water depths of 20 m (for divers); minimum visibility of 0.5–1m; requires low water currents.
Pros	Can cover large areas quickly using standard resources available at spills.	Can be effective in small areas for rapidly definition of a known patch of oil on the bottom; low tech option; has been proven effective for certain spills.	Accurate determination of oil on bottom; verbal and visual description of extent and thickness of oil and spatial variations.
Cons	Only effective in areas with high water clarity; sediment cover will prevent detection over time; ground truthing required.	Samples a very small area, which may not be representative; too slow to be effective over large area; does not indicate quantity of oil on bottom.	Slow; difficult to locate deposits without GPS; decontamination of diving gear can be costly/time consuming.

TECHNOLOGIES AND TECHNIQUES

Bottom Trawls	Photobathymetry	Geophysical/Acoustic Techniques
Fish nets or trawling gear are towed on the bottom for set distance then inspected for presence of oil.	Aerial stereo photography mapping technique used to identify and map underwater features (a realistic scale is 1:10000).	Sonar system that uses the differential density and sound speeds in oil and sediment to detect oil layers on the bottom; a fathometer records a single line under the sounder; side-scan sonar records a swath; output can be enhanced to increase detection.
Readily available in commercial fishing areas.	Available from most private aerial mapping companies, with specifications.	Requirements vary; often not available locally; need trained personnel.
Requires boat and operators to tow the nets; may require multiple vessels to cover large areas; may require many replacement nets as they become oiled.	Aircraft specially equipped to obtain vertical aerial photography with GPS interface.	Requires boat on which equipment can be mounted; requires updated charts so that search area can be defined.
Low coverage; nets have a small sweep area and they have to be pulled up frequently for inspection.	High coverage.	Moderate coverage; data collected at speeds up to m/s.
Quick turnaround.	Slow turnaround.; aerial photographs can be produced in a few days in most places; data interpretation takes one or two additional days.	Medium turnaround; data processing takes hours; preliminary data usually available next day; requires ground truthing.
Low probability; oil staining should be readily differentiated from other fouling materials.	High probability; photography can be used to identify potential sites, which require ground truthing.	High probability; identifies potential sites but all need ground truthing.
Obstructions on the bottom can hang up nets; restricted to relatively shallow depths; sea conditions may restrict vessel operations.	Specifications call for low sun angles and calm sea state; water penetration is limited by water clarity; maximum penetration is 10m for very clear water,1m for turbid water; best if baseline "before" photography is available for comparison.	Sea conditions must be relatively calm to minimize noise in the record.
Can provide data on relative concentrations on the bottom per unit trawl area/time; can survey in grids for more representative areal coverage.	Rapid assessment of large areas; high spatial resolution; good documentation and mapping.	Can be used to identify potential accumulation areas; complete systems can generate high-quality data with track lines, good locational accuracy.
Very slow; nets can fail from excess accumulation of debris.	Limited by water clarity, sun angle, and availability of historic photography for comparisons.	Data processing can be slow; requires extensive ground truthing; requires skilled operators.

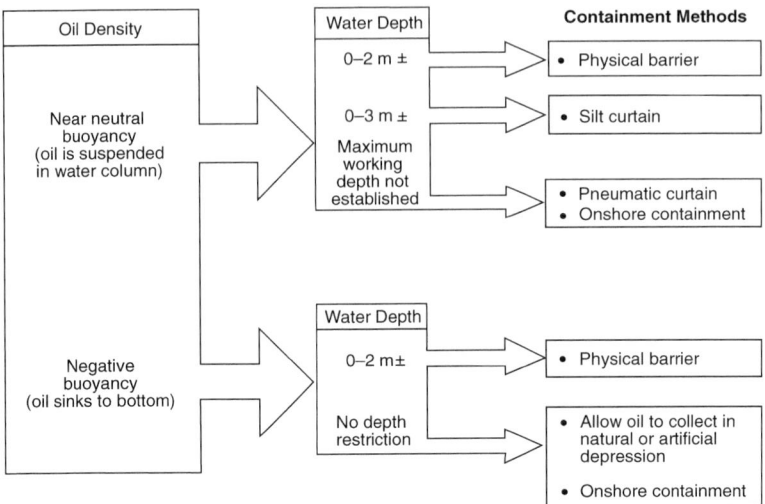

FIGURE 3-2 Decision tree for containment options for sunken oil. Source: Castle et al., 1995.

Identification of natural depressions and collection points, however, may be very useful for locating sedimented oil and planning for its recovery.

Bottom Booms. Bottom-mounted boom systems could be used to contain oil on the seabed. The booms could be moored to the seabed and flotation used to maintain the vertical structure of the boom. These systems are only suitable for locations with low currents and little wave activity. No practical applications of these systems have been reported.

Recovery

The recovery of sunken oil has proven to be very difficult and expensive because the oil is usually widely dispersed. Several of the most widely used recovery methods are reviewed below.

Manual Removal

The manual removal of oil, one of the most widely used recovery methods, involves divers or boat-based personnel using dip nets or seines to collect oil, which is temporarily stored in bags or containers. The purpose of manual recovery is to remove the oil and minimize the collection, handling, treatment, storage, and disposal of other material (oiled sediment, sediment, and water). This approach can be useful for widely dispersed oil, and its effectiveness can be assessed by

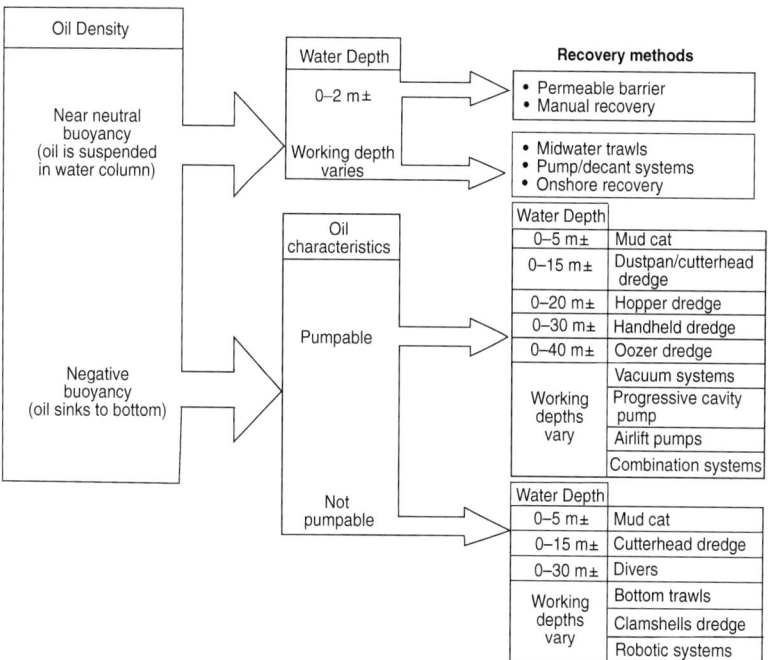

FIGURE 3-3 Decision tree for recovery options for sunken oil. Source: Castle et al., 1995.

cleanup standards or criteria. The biggest disadvantages of manual removal are the large manpower and logistical requirements, slow rates of recovery, strong dependency on weather conditions, and the potential for the oil to be transported while it is being recovered.

Pump and Vacuum Systems

These systems have historically been most successful for removing large volumes of sunken oil. They typically consist of a submersible pump/vacuum system, an oil-water separator, and a storage container. The systems can be mounted on trucks, on land, or on barges or ships. The suction head of the system is normally directed and controlled by divers and may have an air or water injection system to assist in fluidizing and transporting the slurry. The pumped material is usually a mixture of water, oil, and oiled sediment. Highly viscous or solid oils are usually not pumpable and, hence, are not recoverable with this method.

High-energy pumping systems cannot be used because of their potential for breaking up oil droplets or globules and emulsifying the oil. The pumped mixture is typically routed to an oil-water separator from which the oil and oiled sediment are removed and stored. The water may be stored for treatment or released into the sea. Oil-water separation may be difficult if the recovered oil is denser than the recovered water. Pumps and vacuum systems are effective if the oil is localized but are not practical for large areas. They also require extensive equipment and the capacity to handle and treat large volumes of water and sediments.

Nets and Trawls

In addition to containing dispersed oil, nets and trawls can also be used as collection devices (Brown and Goodman, 1987; Delvigne, 1987). This approach is most successful when the relative velocity of the water and the oil collected in the net or trawl is low and the viscosity of the oil is high. The effectiveness decreases as the permeability of the net is reduced and flows are diverted around the net (Delvigne, 1987).

Dredging

Dredging is an efficient, well developed method for removing large volumes of sediment (and oil) from the seabed at high recovery rates. Castle et al. (1995) provide a summary of the operating characteristics of a wide variety of dredging systems routinely considered for the removal of sunken oil. Additional information on the feasibility of dredging for the cleanup of sunken oil is given in Bonham (1989). Large volumes of water, oil, and sediment are typically generated in the dredging process and must be handled, stored, and disposed of as the recovery operation proceeds. Accurate vertical control of the dredge depths is critical to minimizing the amount of dredged material and the amount of clean sediment contaminated with oil as the result of the dredging operation. Operational costs and logistics requirements are lower for land-based than for barge-based methods of handling and storing dredged materials. Given the potential for storms that increase freshwater flows and shipping traffic, both of which can resuspend or remobilize sunken oil, the timeliness of dredging is crucial.

Onshore Recovery

In some cases, oil that has been submerged and mixed with sediment enters the surf zone and is eventually moved onshore and deposited on the shoreline. In these cases, conventional shoreline cleanup methods can be used to remove the oil.

Summary

The containment and recovery of oil dispersed in the water column or deposited on the seabed are very difficult. The problem begins with locating the oil and determining its status. The success of current methods varies greatly but is usually

TABLE 3-3 Options for Containing Oil Suspended in the Water Column

	Pneumatic Barriers	**Net Booms**	**Silt Curtains**
Description	Piping with holes is placed on the bottom, and compressed air is pumped through it, creating an air bubble barrier.	Floating booms with weighted skirts (1-2 m long) composed of mesh designed to allow water to pass while containing suspended oil.	During dredging operations, silt curtains are deployed as a physical barrier to the spread of suspended oil; weighted ballast chains keep the curtain in place.
Availability of Equipment	Uses readily available equipment, although in unique configuration.	There are commercially available net booms have been developed and tested for containing spills of Orimulsion; little availability in the United States.	Not readily available; limited expertise in deployment and maintenance.
Logistical Requirements	Moderate; requires a system to deploy and maintain bubbler; piping has tendency to clog; high installation costs.	Moderate; similar to deployment of standard booms, but with added difficulty because of longer skirt; can become heavy and unmanageable.	Moderate; deployment and maintainance.
Operational Limitations	Only effective in low currents (< 0.2 m/sec), small waves, and shallow water >2 m.	In field tests, the booms failed in currents <0.75 knots; very limited few conditions.	Only effective in very low currents(<10cm/sec); practical limits on curtain depth are 3–6m, which normally doesn't extend to the bottom.
Optimal Conditions	To contain oil spilled in dead-end canals and piers; to protect water intakes.	Will contain oil only in very low-flow areas, such as dead-end canals and piers.	Still water bodies such as lakes; dead-end canals.
Pros	Does not interfere with vessel traffic.	Can be deployed similar to traditional booms.	Can be deployed throughout the entire water column.
Cons	Only effective under very limited conditions; takes time to fabricate and deploy, thus only effective where pre-deployed; little data available to assess performance.	Only contains oil suspended in the upper water column, to the depth of the mesh skirt; unknown whether the mesh will clog and fail at lower currents.	Effective under very limited conditions, not likely to coincide with location where oil needs containment; oil droplets are larger than silt and could clog curtain.

TABLE 3-4 Options for Recovering Oil Deposited on the Seabed

	Manual Removal by Divers	**Nets/Trawls**
Description	Divers pick up solid and semi-solid oil by hand or with nets on the bottom, placing it in bags or other containers	Fish nets and trawls are dragged on the bottom to collect solidified oil
Equipment Availability	Contaminated-water dive gear may not be locally available	Nets and vessels readily available in areas with commercial fishing industry
Logistical Needs	Moderate; diving in contaminated water requires special gear and decon procedures; handling of oily wastes on water can be difficult	Low; uses standard equipment, though nets will have to be replaced often because of fouling
Operational Limitations	Water depths up to 60-80 ft for routine dive operations; water visibility of 1-2 ft so divers can see the oil; bad weather can shut down operations	Water depths normally reached by bottom trawlers; obstructions on the bottom which will hang up nets; rough sea conditions; too shallow for boat operations
Optimal Conditions	Shallow, protected areas where dive operations can be conducted safely; small amount of oil; scattered oil deposits	Areas where bottom trawlers normally work; solidified oil
Pros	Divers can be very selective, removing only oil, minimizing the volume of recovered materials; most effective method for widely scattered oil deposits	Uses available resources; low tech
Cons	Large manpower and logistics requirements; problems with contaminated water diving and equipment decon; slow recovery rates; weather dependent operations	Not effective for liquid or semi-solid oil; nets can quickly become clogged and fail; can become heavy and unmanageable if loaded with oil; could require many nets which are expensive

limited because the oil, which is mixed with sediments and water, is usually widely dispensed. In general, the success is greatest when the current speeds and wave conditions at the spill site are low, the oil is pumpable, the water depths are relatively shallow, and the sunken oil has concentrated in depressions or collection areas. The selection of containment and recovery methods is highly dependent on the specific location and environmental conditions during the spill, the

TECHNOLOGIES AND TECHNIQUES *51*

Pump and Vacuum Systems (Diver-directed)	**Dredging**
Divers direct a suction hose connected to a pump and vacuum system, connected to oil-water separator, and solids containers. Viscous oils require special pumps and suction heads. Even in low water visibility, divers can identify oil by feel or get feedback from top-side monitors of changes in oil recovery rates in effluents	Special purpose dredges, usually small and mobile, with ability for accurate vertical control. Uses land or barge-based systems for storage and separation of the large volumes of oil-water-solids.
Readily available equipment but needs modification to spill conditions, particularly pumping systems, and capacity for handling large volumes of materials during oil-water-solids separation	Varies; readily available in active port areas; takes days/week to mobilize complete systems
High, especially if recovery operations are not very close to shore. On-water systems will be very complicated and subject to weather, vessel traffic, and other safety issues.	High, especially if recovery operations are not very close to shore, because of large volumes of materials handled. On-water systems will be very complicated and subject to weather, vessel traffic, and other safety issues.
Water depths up to 60-80 ft for routine dive operations; water visibility of 1-2 ft so divers can see the oil; bad weather can shut down operations; solid oil which is not pumpable	Min/max water depths are a function of dredge type, usually 2-100 ft; not in rocky substrates; bad weather can shut down operations
Sites adjacent to shore, requiring minimal on-water systems; liquid or semi-solid oil; thick oil deposits, good visibility; low currents	Large volume of thick oil on the bottom; need for rapid removal before conditions change and oil is remobilized, buried by clean sediment, or will have larger environmental effects
Most experience is with this type of recovery; diver can be selective in recovering only oil and effective with scattered deposits;	Rapid removal rates; can recover non-pumpable oil
Very large manpower and logistics requirements, including large volumes of water-oil-solids handling, separation, storage, and disposal; problems with contaminated water diving and equipment decon; slow recovery rates; weather dependent operations	Generates large volumes of water/solids for handling, treatment, disposal; large logistics requirements; could re-suspend oil/turbidity and affect other resources

characteristics of the oil and its state of weathering and interaction with sediments, the availability of equipment, and logistical support for the cleanup operation. In addition, the potential environmental impacts of implementing these methods, particularly in sensitive benthic habitats, must be considered. Tables 3-3 and 3-4 summarize the uses and limitations of various containment and recovery methods.

4

Barriers to Effective Response

In presentations at the four committee meetings and at the workshop, leading experts from the spill-response, regulatory, environmental, and oil-transportation communities consistently identified a number of barriers to effective responses to spills of nonfloating oils. The major managerial, technological, and financial barriers identified by these experts and supported by the experience of committee members are summarized below.

MANAGERIAL BARRIERS

A major managerial barrier in responding to spills of nonfloating oils is the lack of experience at the local level. The knowledge base for planning and responding to oil spills is primarily derived from responses to actual oil spills. Significant oil spills are infrequent by their very nature, and spills of nonfloating oils are only a small fraction of all oil spills. Thus, it is difficult to acquire and maintain a sufficient knowledge base at the local level to respond to nonfloating-oil spills, particularly because few organizations have full-time, dedicated response teams. Furthermore, planning for nonfloating-oil spills generally has a low priority because of their infrequency. Responding to a spill of nonfloating oils is, therefore, often a new or very rare experience for local response teams who are likely to have trouble anticipating problems and formulating effective response strategies.

Planning for spills of nonfloating oils at the regional level has often been inadequate. There are 44 area committees in the USCG's jurisdiction. None of the area plans, however, has a well developed strategy for responding to spills of nonfloating oils. As a result, planners have limited experience in identifying the likelihood and potential sources of spills of nonfloating oils, determining resources at risk, establishing protection priorities and strategies, or evaluating response capabilities in federal, state, and industry plans.

Area committees and other constituencies have not adequately resolved emergency regulatory issues associated with responses to nonfloating-oil spills, such as obtaining permits for emergency dredging and the discharge of co-collected water. As a result, even though every spill of nonfloating oils is a true emergency, difficult regulatory issues must be faced without the benefit of prior discussions of response options. Consequently, regulatory agencies cannot usually provide timely approvals.

The resources and information necessary to respond effectively to nonfloating-oil spills have not been identified, including divers capable of operating in contaminated waters, the capability of updating bathymetric maps to determine potential accumulation zones, and the selection and implementation of systems to track the movement and distribution of subsurface oil. Furthermore, few, if any, drills or exercises have been carried out with scenarios focused on spills of nonfloating oils. In the absence of a real spill, exercises are an excellent mechanism for verifying response plans and improving response capabilities. The lack of drills, combined with limited experience with actual spills, has seriously impeded the development of a practical knowledge base for responders.

Misconceptions about the behavior, fate, and effects of nonfloating-oil spills are widespread. Descriptions of the transport and fate of spills of nonfloating oils have been confused and inconsistent. Consequently, the documentation of actual spills is poor and difficult to interpret, and no formal system for sharing lessons learned from previous spills has been developed. Without field experience or adequate literature on which to base predictions of behavior and effects, resource managers and responders have been forced to develop their own conceptual models of how nonfloating oils might behave and their environmental impacts. These conceptual models are often inadequate or incorrect, leading to erroneous assumptions about the viability or effectiveness of response options.

TECHNOLOGICAL BARRIERS

Existing methods for tracking spills are not effective for tracking nonfloating oils. One of the first questions asked after an oil spill is where the oil is going. The answer to this question often determines subsequent decisions. Most conventional methods for predicting the trajectory and tracking oil spills rely on two-dimensional (e.g., surface) transport and fate models and visual observations, none of which is effective for tracking nonfloating oils.

Methods used to track nonfloating oils in past spills have been largely ineffective. Most existing methods have low encounter rates and limited areal coverage for tracking oil suspended in the water column. Thus, it is impossible to generate a synoptic map of the dispersed oil plume over time. The problems are similar for tracking oil deposited on the seabed, but generally the movement of deposited oil is less dynamic. Bottom sampling methods for large areal searches (e.g., video and sonar searches) are limited by site constraints, difficult logistics, and the need for extensive ground truthing (i.e., *in situ* verification). The most commonly used techniques (e.g., sorbent drops and drags, diver observations, bottom trawls) can only sample limited areas and are slow, labor intensive, and logistics intensive.

The options for effectively containing and recovering nonfloating oils are limited. Even the most promising methods have not been effective for containing and recovering oils mixed in the water column, except under ideal conditions (e.g., small spills of emulsified oils in areas with very low currents and little wave activity). Generally, oil in the water column disperses quickly over large areas and volumes, becoming unavailable for effective recovery. Containment of oil deposited on the seabed is only feasible where the oil accumulates naturally. In these cases, recovery rates can be very high with the use of manual, pumping, or dredging techniques. However, each of these methods requires handling large volumes of water and solids.

Because of a general lack of knowledge about benthic habitats and resources, assessing resources at risk from nonfloating oils is extremely difficult. Area plans include annexes, in which sensitive areas are identified and prioritized for protection. One of the tasks of area committees is to discuss cleanup methods and end points appropriate for different habitats. Although nonfloating-oil spills threaten both the water-column and bottom (benthic) habitats, data on benthic habitats and resources at risk are either very sparse or not available. Benthic habitats are often described in very general terms, and few areas have been mapped in detail. Areas with high concentrations of plant or animal species or sites important to the sensitive, early life stages of organisms are usually poorly known, even for species with high commercial value. Without this information, it is difficult for resource managers to evaluate the potential effects of unrecovered oil or to decide on how aggressive their containment and recovery efforts should be.

FINANCIAL BARRIERS

Funding levels for testing and evaluating potential response options for all oil spills are low, but they are especially low for spills of nonfloating oils (NRC, 1998). Even after the watershed *Exxon Valdez* oil spill, federal, state, and industry funding for research and development have remained low. Funding for research and development on the containment, recovery, and effects of nonfloating oils has also been low and is generally targeted toward emulsified fuels for which funding is provided by the producers of these fuels. The lack of research, development, testing, and evaluation has left responders with a very limited number of unproven options for responding to nonfloating-oil spills. Information about how these options might be used under specific spill conditions is also limited.

5

Findings, Conclusions, and Recommendations

FINDINGS

Finding 1. From 1991 to 1996, approximately 17 percent of the petroleum products transported over U.S. waters were heavy oils and heavy-oil products, such as residual fuel oils, coke, and asphalt. Approximately 44 percent was moved by barge and 56 percent by tanker.

Finding 2. From 1991 to 1996, approximately 23 percent of the petroleum products spilled in U.S. waters were heavy oils. In only 20 percent of these spills did a significant portion of the spilled products sink or become suspended in the water column. Most of the time, spills of heavy oil remained on the surface. The average number of spills of more than 20 barrels of heavy oil and asphalt was 16 per year, with an average volume of 785 barrels per spill. The committee projects that a 30 percent reduction in the number and volume of heavy-oil spills would have been realized if tankers and barges had all been double-hulled vessels.

Finding 3. In recent years, barges have had significantly higher spill rates than tankers. From 1991 to 1996, barges accounted for approximately 80 percent of the volume of heavy-oil spills, and the spill rate, expressed in terms of barrels-spilled-per-ton-mile, was more than 10 times higher for barges than for tankers. Although the reduction in spill volume from tank barges since 1990 has been significant (about one-third of pre-1990 volume), the reduction for tankers has been even more dramatic (about one-tenth of pre-1990 volume).

Finding 4. Specific gravity, as used in the regulatory definition of Group V oils, does not adequately characterize all oil types and weathering conditions that produce nonfloating oils. The committee was asked to address the issue of responses to Group V oil spills, defined by current regulations as oils with a specific gravity of greater than 1.0. However, the committee determined that the issue of concern is planning for and responding to oil spills in which most, or a significant quantity, of the spilled oil does not float. The committee, therefore, decided to use the term "nonfloating oils" to describe the oils of concern.

Finding 5. Nonfloating oils behave differently and have different environmental fates and effects than floating oils. The resources at greatest risk from spills of floating oils are those that use the water surface and the shoreline. Floating-oil spills seldom have significant impacts on water-column and benthic resources. In contrast, nonfloating-oil spills pose a substantial threat to water-column and benthic resources, particularly where significant amounts of oil have accumulated on the seafloor. Nonfloating oils tend to weather slowly and thus can affect resources for long periods of time and at great distances from the release site. However, the effects and behavior of nonfloating oil are poorly understood.

Finding 6. Although spill modeling and supporting information systems are well developed, they are not commonly used in response to nonfloating-oil spills because of limited environmental data and observations of oil suspended in the water or deposited on the seabed. Oil-spill models and supporting information systems are routinely used in contingency planning and spill responses. Sophisticated, user-friendly interfaces have been developed to take advantage of the latest advances in computer hardware and software. The current generation of models can rapidly incorporate environmental data from a variety of sources and include integrated geographic information systems. The models can also assimilate data on the most recently observed location of spilled oil and have improved forecasts of oil movements. They are not routinely used, however, in response to nonfloating-oil spills because of the lack of supporting data on the three-dimensional currents and concentrations of suspended sediments. Field data, such as oil concentrations in the water column and on the seabed, are also not generally available to validate or update models.

Finding 7. A substantial number of techniques and tools for tracking subsurface oil have been developed. Most of them, however, have not been used in response to actual oil spills. Many techniques are available for determining the location of oil both in the water column and on the seabed. These include visual observations, geophysical and acoustic methods, remote sensing, water-column and seabed sampling, *in situ* detectors, and nets and trawl sampling. The most direct and simplest methods, such as diver observations and direct sampling, are widely used, but they are labor intensive and slow. More sophisticated approaches, such as remote sensing, are limited to zones very near the sea surface because of technical constraints. Other advanced technologies, such as acoustic techniques, cannot differentiate between oil and water or between oiled sediments and underlying sediments. Many of the more sophisticated systems are prone to misuse and produce ambiguous data that are subject to misinterpretation. The performance of all but the simplest methods is undocumented either by field experiments or by use in spill responses.

Finding 8. Although many technologies are available for containing and recovering subsurface oil, few are effective, and most work only in very limited environmental conditions. Containment of oil suspended in the water column using silt curtains, pneumatic barriers, and nets and trawls is only effective in areas with very low currents and minimal wave activity. These conditions rarely exist at spill sites, particularly at sites in estuarine or coastal waters. The recovery of oil in the water column by trawls and nets is limited by the viscosity of the oil and net tow speeds.

The containment of oil on the seabed is typically ineffective, except at natural collection points (e.g., depressions and areas of convergence). The collection of oil on the seabed by manual methods, in natural collection areas and along the shoreline after beaching, is effective but labor intensive and slow. Manual methods are also limited by the depths at which diver-based operations can be carried out safely. Dredging techniques have rarely been used because of limited recovery rates, the large volumes of water and sediment generated, and the problems of storing, treating, and discharging co-produced materials.

Finding 9. The lack of knowledge and lack of experience, especially at the local level, in responding to spills of nonfloating oils is a significant barrier to effective response. The knowledge base and response capabilities for tracking, containing, and recovering nonfloating oils have not been adequately developed. Even at the national level, no system has been developed for sharing experiences or documenting the effectiveness and limitations of various options. With limited experience and a lack of proven, specialized systems, responders have found it difficult to adapt available equipment for responses to spills of nonfloating oils.

Finding 10. Planning for spills of nonfloating oils is inadequate at the local level. Existing area contingency plans do not include comprehensive sections on the risk of spills of nonfloating oils or how to respond to them. To date, planning has focused primarily on spills of floating oils. Inventories of equipment, lists of specialized services, assessments of the resources at risk, and protection priorities have not been developed by area committees for nonfloating oils. Nor have they identified the risks (e.g., transportation patterns, volumes, oil types), developed appropriate scenarios and response plans, or reviewed acceptable cleanup methods and end points. Existing plans have not been tested during drills or exercises to address deficiencies.

Finding 11. Funding levels for research, development, testing, and evaluation of spills of nonfloating oils are very low. The only active research programs currently under way either by government or industry groups are focused on emulsified fuel oils. Because the risk of spills of nonfloating oils is perceived as low relative to spills of floating oils, few research and development funds have been committed.

CONCLUSIONS

Conclusion 1. The tracking, containment, and recovery of spills of nonfloating oils pose challenging problems, principally because nonfloating oils suspended in the water column become mixed with large volumes of seawater and may interact with sediments in the water column or on the seabed. The ability to track, contain, and recover nonfloating oils is critically dependent on the physical and chemical properties of the oils and the water or the oils and the other materials dispersed in the water column or on the seabed. The differences in these characteristics are often quite small, and little technology is available for determining them.

Conclusion 2. Although many methods are available for tracking nonfloating oils, the simplest and most reliable are labor intensive and cover only limited areas. More sophisticated methods have severe technical limitations, require specialized equipment and highly skilled operators, or cannot distinguish oil from water or other materials dispersed in the water column. Engineered systems for containing oil in the water column or on the seabed are few and only work in environments with low currents and minimal waves. Natural containment in seabed depressions or in the lee of topographical or man-made structures on the seabed is effective for containing oils, but these are not always available in the vicinity of the spill.

Conclusion 3. The recovery of oil from the water column is very difficult because of the low concentration of dispersed oil; hence, recovery is rarely attempted. If oil collects on the seabed in natural containment areas, many options for effective recovery are available, although most of them are labor intensive and access to response equipment is a problem.

Conclusion 4. The volume and frequency of spills of nonfloating oils is significant (although smaller than for floating oils) and, therefore, should be an integral part of planning for spill responses, particularly in areas where nonfloating oils are regularly transported. Transport by tank barges raises particular concerns, given the relatively high spill rates from these vessels. The risks of potential harm to water-column and benthic resources from nonfloating oils have not been adequately addressed in the contingency plans for individual facilities or geographic areas.

Conclusion 5. Inland barges are subject to greater risks of spills than tankers and coastal barges; consequently, spill rates for barges are likely to be higher than for tankers. However, the large difference between the overall spill rates, as well as the decreasing number of spills from tankers in recent years (post-OPA 90), raises concerns regarding the performance of barges.

RECOMMENDATIONS

The recommendations below are intended to improve the capability of the spill response community to respond to spills of nonfloating oils.

Recommendation 1. The U.S. Coast Guard should direct area planning committees to assess the risk of spills of nonfloating oils (i.e., oils that may be dispersed in the water column or ultimately sink to the seabed) to determine the resources at risk. In areas with significant environmental resources risk, area planning committees should develop response plans that include consultation and coordination protocols and should obtain pre-approvals and authorizations to facilitate responses to spills. Stakeholder groups should be educated about the impact and methods available for tracking, containing, and recovering oil suspended in the water column or on the seabed. Area committees in locations where there is a high risk of spills of nonfloating oils should include at least one scenario for responding to a nonfloating-oil spill in their training or drill programs.

Recommendation 2. The U.S. Coast Guard should improve its knowledge base, education, and training for responding to spills of nonfloating oils by including a scenario involving a spill of nonfloating oils in oil-spill response drills, by establishing a knowledge base and scientific support teams to respond to these types of spills, and by disseminating this knowledge to the federal spill-response coordinators and area planning committees as part of ongoing training programs. The information would help area planners assess the requirements for responding to nonfloating-oil spills.

Recommendation 3. The U.S. Coast Guard should support the development and implementation of an evaluation program for tracking oil in the water column and on the seabed, as well as containment and recovery techniques for use on the seabed. The findings of these evaluations should be documented and distributed to the environmental response community to improve response plans for spills of nonfloating oils.

Recommendation 4. Tests of area contingency plans and industry response plans for responses to spills of nonfloating oils should be required parts of training and drill programs.

Recommendation 5. The U.S. Coast Guard should monitor spill rates from tank barges to ascertain whether current regulatory requirements and voluntary programs will reduce the frequency and volume of spill incidents. If not, the Coast Guard should consider initiating regulatory changes.

References

Alejandro, A.C., and J.L. Buri. 1987. *M/V Alvenus*: Anatomy of a major oil spill. Pp. 27–32 in Proceedings of the 1987 Oil Spill Conference. Washington, D.C.: American Petroleum Institute.

Anderson, E.L., E. Howlett, K. Jayko, V. Kolluru, M. Reed, and M. Spaulding. 1993. The worldwide oil spill model (WOSM): an overview. Pp. 627–646 in Proceedings of the 16th Arctic and Marine Oil Spill Program, Technical Seminar. Ottawa, Ontario: Environment Canada.

Anderson, E., C. Galgan, and E. Howlett. 1998. The on-scene command and control system (OSC 2): an integrated incident command system forms-database management system and oil spill trajectory and fates model. Pp. 449–463 in Proceedings of the 21st Arctic and Marine Oil Spill Program, Technical Seminar. Ottawa, Ontario: Environment Canada.

Anderson, J.W., R. Riley, S. Kiesser, and J. Gurtisen. 1987. Toxicity of dispersed and undispersed Prudhoe Bay crude oil fractions to shrimp and fish. Pp. 235–240 in Proceedings of the 1987 Oil Spill Conference. Washington, D.C.: American Petroleum Institute.

ASA (Applied Science Associates, Inc). 1997. OILMAP technical and user's manual. Narragansett, R.I.: Applied Science Assoicates, Inc.

ASCE (American Society of Civil Engineers) Task Committee on Modeling of Oil Spills. 1996. State of the art review of modeling transport and fate of oil spills. Journal of Hydraulic Engineering 22(11): 594–609.

Benggio, B.L. 1994a. Diaper drop technique for locating submerged oil, Presented at Group V Working Group Meeting, December 15, 1994, in NOAA 1997.

Benggio, B.L. 1994b. Photobathymetry for submerged oil, Presented at Group V Working Group Meeting, December 15, 1994, in NOAA 1997.

Benggio, B.L. 1994c. An evaluation of options for removing submerged oil offshore Treasure Island, Tampa Bay Oil Spill. Report HMRAD 94-5 NOAA. Seattle, Wash.: Hazardous Materials Response and Assessment Division. NOAA.

Bonham, N. 1989. Response techniques for the cleanup of sinking hazardous materials, Report EPS 4/SP/1. Ottawa, Ontario: Environment Canada.

Boyer, K. R., V.E. Hodge, and R.S. Wetzel. 1987. Handbook: Responding to Discharges of Sinking Hazardous Substances. Science Applications International Corporation (SAIC), EPA, September 1987, Environmental Protection Agency.

Brown, C.E. 1998. Laser induced fluorescence study of Orimulsion. Final report, April 30, 1998. Ottawa, Ontario: Emergencies Science Division, Environment Canada.

Brown, H., E.H. Owens, and M. Green. 1997. Submerged and sunken oil: behavior, response options, feasibility, and expectations. Pp. 135–146 in Proceedings of the 21st Arctic and Marine Oil Spill Program, Technical Seminar. Ottawa, Ontario: Environment Canada.

Brown, H., and R.H. Goodman. 1987. The recovery of spilled heavy oil with fish netting. Pp. 123–126 in Proceedings of the 1989 International Oil Spill Conference. Washington, D.C.: American Petroleum Institute.

Burns, G.H., C.A. Benson, T. Eason, S. Kelly, B. Benggio, J. Michel, and M. Ploen. 1995. Recovery of submerged oil at San Juan, Puerto Rico, 1994. Pp. 551–557 in Proceedings of the 1995 Oil Spill Conference. Washington, D.C.: American Petroleum Institute.

Castle, R.W., F. Wehrenburg, J. Bartlett, and J. Nuckols. 1995. Heavy oil spills: out of sight, out of mind. Pp. 565–571 in Proceedings of the 1995 International Oil Spill Conference. Washington, D.C.: American Petroleum Institute.

Chivers, R.C., N. Emerson, and D.R. Burns. 1990. New acoustic processing for underwater surveying. Hydrographic Journal 56:9-16.

Deis, D.R., N.G. Tavel, P. Masciangioli, C. Villoria, M.A. Jones, G.F. Ortega, and G.R. Lee. 1997. Orimulsion: research and testing and open water containment and recovery trials. Pp. 459–467 in Proceedings of the 1997 International Oil Spill Conference. Washington, D.C.: American Petroleum Institute.

Delvigne, G.A.L. 1987. Netting of viscous oils. Pp. 115–122 in the Proceedings of the 1987 International Oil Spill Conference. Washington, D.C.: American Petroleum Institute.

Dick, R., M. Fruhwirth, M. Fingas, and C. Brown. 1992. Laser fluorosensor work in Canada. Pp. 223–236 in Proceedings of the First Thematic Conference: Remote Sensing for Marine and Coastal Environments: Needs and Solutions for Pollution Monitoring, Control and Abatement. Ann Arbor, Mich.: The Environmental Research Institute of Michigan.

Dick, R., and M.F. Fingas. 1992. First results of airborne trials of a 64-channel laser fluorosensor for oil detection. Pp. 365–379 in Proceedings of the 15th Arctic and Marine Oil Spill Program, Technical Seminar. Ottawa, Ontario: Environment Canada.

DOE (U.S. Department of Energy). 1998. Annual Energy Outlook 1999 (early release). Washington, D.C.: DOE.

Elliot, A.J. 1991. EUROSPILL: Oceanographic processes and NW European shelf databases. Marine Pollution Bulletin 22: 548–553.

Federal Register. 1996. 61(4): 7921, Feb. 29, 1996.

Fingas, M.F., and C.E. Brown. 1996. Review of oil spill remote sensing. Pp. 223–236 in the Proceedings of the Ecoinforma '96, Global Networks for Environmental Information. Ann Arbor, Mich.: Environmental Research Institute.

French, D.P., E. Howlett, and D. Mendelsohn. 1994. Oil and chemical impact model system description and application. Pp. 767–784 in the Proceedings of the 17th Arctic and Marine Oil Spill Program, Technical Seminar. Ottawa, Ontario: Environment Canada.

Galagan, C., E. Howlett, and A.J. Brown. 1992. PC-based visualization of geographically referenced environmental data. Pp. 23–29 in Proceedings of the 15th Arctic and Marine Oil Spill Program, Technical Seminar. Ottawa, Ontario: Environment Canada.

Galt, J.A. 1994. Trajectory analysis for oil spills. Journal of Advanced Marine Technology Conference 11: 91–126.

Galt, J.A. 1995. The integration of trajectory models and analysis into spill response information systems. Pp. 499-507 in Proceedings of 2nd International Spill Research and Development Forum. London, U.K.: International Maritime Organization.

Galt, J.A., D.L. Payton, H. Norris, and C. Freil. 1996. Digital distribution standard for NOAA trajectory analysis information. HAZMAT Report 96-4. Seattle, Wash.: Hazardous Materials Response and Assessment Division, NOAA.

Gundlach, E.R., K.J. Finklestein, and J.L. Sadd. 1981. Impact and persistence of Ixtoc I oil on the south Texas coast. Pp. 477–485 in Proceedings of the 1981 Oil Spill Conference. Washington, D.C.: American Petroleum Institute.

Harper, J., G. Sergy, and T. Sagayama. 1995. Subsurface oil in coarse sediments experiments (SOCSEX II). Pp. 867–886 in Proceedings of the 18th Arctic and Marine Oilspill Program Technical Conference. Edmonton, Alberta.: Environment Canada.

Hutchison, J.H., and B.L. Simonsen. 1979. Cleanup operations after the 1976 SS *Sansinena* explosion: an industrial perspective. Pp. 429–433 in Proceedings of the 1979 Oil Spill Conference. Washington, D.C.: American Petroleum Institute.

Johansen, O. 1985. Particle in fluid model for simulation of oil drift and spread. Part I. Basic concepts. Note: 02.0706.40/2/85. Trondheim, Norway: Oceanographic Center, SINTEF.

Jokuty, P., S. Whiticar, M. Fingas, Z. Wang, K. Doe, D. Kyle, P. Lambert, and B. Fieldhouse. 1995. Orimulsion: Physical Properties, Chemical Composition, Dispensability, and Toxicity. Final Report. Ottawa, Ontario: Emergency Services Division, Environment Canada.

Kana, T.W., 1979. Suspended sediment in breaking waves. Department of Geology, University of South Carolina, Columbia, S.C. Unpublished dissertation.

Kennedy, D.M., and B.J. Baca, eds. 1984. Fate and effects of the *Mobiloil* spill in the Columbia River. Ocean Assessments Division, Seattle, Wash.: NOAA.

Kirstein, B., J.R. Clayton, C. Clary, J.R. Payne, D. McNabb, G. Fauna, and R. Redding. 1985. Integration of Suspended Particulate Matter and Oil Transportation Study. Anchorage, Alaska: Mineral Management Service.

Kolluru, V., M.L. Spaulding, and E. Anderson. 1994. A three-dimensional subsurface oil dispersion model using a particle-based approach. Pp. 767–784 in Proceedings of the 17th Arctic and Marine Oil Spill Program, Technical Seminar. Ottawa, Ontario: Environment Canada.

Lee, S.C., D. Mackay, F. Bonville, E. Joner, and W.Y. Shui. 1989. A Study of the Long-Term Weathering of Submerged and Overwashed Oil. EE-119. Ottawa, Ontario: Environment Canada.

Lee, S.C., W.Y. Shui, and D. Mackay. 1992. A Study of the Long-Term Fate and Behavior of Heavy Oils. EE-128. Ottawa, Ontario: Environment Canada.

Marine Microsystems, Inc. 1992. ROXANN Sea trails, T/V Haven, Gulf of Genoa. City, State: Marine Microsystems, Inc.

Markarian, R.K., J.P. Nicolette, T.R. Barber, and L.H. Giese. 1993. A critical review of toxicity values and evaluation of the persistence of petroleum products for use in natural resource damage assessments. Washington, D.C.: American Petroleum Institute.

Martinelli, M., A. Luise, E. Tromellini, T.C. Sauer, J.M. Neff, and G.S. Douglas. 1995. *The M/C Haven* oil spill: environmental assessment of pathways and resource injury. Pp. 679–685 in the Proceedings of the 1995 Oil Spill Conference. Washington, D.C.: American Petroleum Institute.

McCourt, J., and L. Shier. 1998. Interaction between oil and suspended particulate matter in the Yukon River. Pp. 79–87 in Proceedings of the 21st Arctic and Marine Oil Spill Program, Technical Conference. Edmonton, Alberta: Environment Canada.

Michel, J., and J. Galt. 1995. Conditions under which floating oil slicks can sink in marine settings. Pp. 573–576 in Proceedings of the 1995 International Oil Spill Conference. Washington, D.C.: American Petroleum Institute.

Michel, J., D. Scholz, C.B. Henry, and B.L. Benggio. 1995. Group V Fuel Oils: Source, Behavior, and Response Issues. Pp. 559–564 in Proceedings of the 1995 International Oil Spill Conference. Washington, D.C.: American Petroleum Institute.

MMS (Minerals Management Service). 1998. Oil Spills 1991–1997: Statistical Report. Herndon, Va.: U.S. Department of the Interior.

Moller, T.H. 1992. Recent experience of oil sinking. Pp. 11–14 in Proceedings of the 15th Arctic and Marine Oil Spill Program, Technical Seminar. Ottawa, Ontario: Environment Canada.

Nielsen, P. 1992. Coastal Bottom Boundary Layers and Sediment Transport. Singapore: Scientific Press.
NOAA (National Oceanic and Atmospheric Administration).. 1992. Oil Spill Case Histories, 1967–1991. Report No. HMRAD 92-11. Seattle, Wash.: Hazardous Materials Response and Assessment Division, NOAA.
NOAA. 1995. Barge *Morris J. Berman* spill: NOAA's Scientific Response. HAZMAT Report No. 95-10. Seattle, Wash.: Hazardous Materials Response and Assessment Division, NOAA.
NOAA. 1997. Oil beneath the water surface and review of currently available literature on Group V oils: an annotated bibliography. Report HMRAD 95-8. January 1997 update. Seattle, Wash.: Hazardous Materials Response and Assessment Division, NOAA.
NOAA and API (American Petroleum Institute). 1995. Environmental considerations for oil spill response. Seattle, Wash.: Hazardous Materials Response and Assessment Division, NOAA.
NRC (National Research Council). 1998. Double-Hull Tanker Legislation: An Assessment of the Oil Pollution Act of 1990. Washington, D.C.: National Academy Press.
Ostazeski, S.A., S.C. Macomber, L.G. Roberts, A.D. Uhler, K.R. Bitting, and R. Hiltabrand. 1997. The environmental behavior of Orimulsion^{TM} spilled on water. Pp. 469–477 in Proceedings of the 1997 Oil Spill Conference. Washington, D.C.: American Petroleum Institute.
OTA (Office of Technology Assessment). 1990. Coping with an Oiled Sea: an Analysis of Spill Response Technologies. Background Paper. OTA-BP-O-63. Washington, D.C.: Government Printing Office.
Ploen, M. 1995. Submerged oil recovery. Pp. 165–173 in Proceedings of the 2nd International Oil Spill Research and Development Forum. London, U.K.: International Maritime Organization.
Proctor, R., A.J. Elliot, and R.A. Flather. 1994. Forecast and simulations of the *Braer* oil spill. Marine Pollution Bulletin 28: 219–229.
Rains, G. Subject Presentation by Gloria Rains, representative of ManaSota-88 Inc., to the Committee on Marine Transportation of Heavy Oils. Washington, D.C., August 20, 1998.
Reed, M., E. Gundlach, and T. Kana. 1989. A coastal zone oil spill model: development and sensitivity studies. Oil and Chemical Pollution 5: 411–449.
Scholz, D.K., J. Michel, C.B. Henry, Jr., and B. Benggio. 1994. Assessment of risks associated with the shipment and transfer of Group V fuel oils. HAZMAT Report No. 94-8. Seattle, Wash.: Hazardous Materials Response and Assessment Division, NOAA.
Smedley, J.B., and R.C. Belore. 1991. Review of possible technologies for the detection and tracking of submerged oil. TP 10787E. Montreal, Quebec: Transport Development Center, Transport Canada.
Sommerville, M., T. Lunel, N. Bailey, D. Oaland, C. Miles, P. Gunter, and T. Waldhoff. 1997. Orimulsion. Pp. 479–467 in Proceedings of the 1997 International Oil Spill Conference. Washington, D.C.: American Petroleum Institute.
Spaulding, M.L. 1995. Oil spill trajectory and fate modeling: a state of the art review. P. 55 in the Proceedings of the 2nd International Oil Spill Research and Development Forum. London: International Maritime Organization.
Spaulding, M.L., A. Modulo, and V.S. Kolluru. 1992. A hybrid model to predict the entrainment and subsurface transport of oil. Pp. 67–92 in the Proceedings of the 15th Arctic and Marine Oil Spill Program, Technical Seminar. Ottawa, Ontario: Environmental Canada.
Spaulding, M.L., V.S. Kolluru, E. Anderson, and E. Howlett. 1994. Application of three dimensional oil spill model (WOSM/OILMAP) to hindcast the Braer spill. Spill Science and Technology Bulletin 1(1): 23–35.
Spaulding, M.L. and A. Chen. 1994. A shell-based approach to worldwide oil spill modeling. Journal of Advanced Marine Technology 11: 127–142.
Turner Designs. 1999. Available on line at HtmlResAnchor www.uturnerdesigns.com/oilinwater.

REFERENCES

USACE (U.S. Army Corps of Engineers). 1998a. Waterborne Commerce of the United States, Calendar Year 1996. Part 5. Waterways and Harbors National Summaries. New Orleans, La.: USACE.

USACE. 1998b. Summary of Foreign and Domestic Waterborne Commerce of Petroleum and Petroleum Products for 1990–1996. Prepared for the Committee on the Marine Transportation of Heavy Oil by the U.S. Army Corps of Engineers. New Orleans, La.: USACE.

USCG (U.S. Coast Guard). 1998. Spill Data for Oil Spills in U.S. Waters. Washington, D.C.: USCG.

Vincente, V. 1994. Field survey report, *Morris J. Berman* oil spill. San Juan, Puerto Rico: National Marine Fisheries Service.

Weems, L.H., I. Byron, J. O'Brien, D.W. Oge, and R. Lanier. 1997. Recovery of Lapio from the bottom of the lower Mississippi River. Pp. 773–776 in Proceedings of the 1997 Oil Spill Conference. Washington, D.C.: American Petroleum Institute.

Yaroch, G.N., and G.A. Reiter. 1989. The tank barge MCN-5: lessons in salvage and response guidelines. Pp. 87–90 in Proceedings of the 1989 Oil Spill Conference. Washington, D.C.: American Petroleum Institute.

Appendices

APPENDIX A

Biographical Sketches of Committee Members

Malcolm L. Spaulding (chair) is professor and chair of the Department of Ocean Engineering at the University of Rhode Island, where he has been a member of the faculty since 1973. He is an expert in numerical modeling of nearshore and coastal processes, including hydrodynamics, oil and pollutant transport and fate, waves, and sediment transport. Dr. Spaulding received his Ph.D. in mechanical engineering and applied mechanics from the University of Rhode Island. He serves on a number of national and international research and advisory organizations related to coastal and ocean processes. He has served on National Research Council committees for the Marine Board and the Ocean Studies Board and is currently a member of the Marine Board.

Malcolm MacKinnon III (NAE) retired in 1990 from the U.S. Navy, where he was the chief engineer of the Navy and the vice commander, Naval Sea Systems Command. RADM MacKinnon is the president and chief executive officer of MSCL, Inc., a consulting firm that provides technical services to the maritime industry (military and commercial) worldwide. He is a graduate of the U.S. Naval Academy and holds advanced degrees in naval architecture and marine engineering from the Massachusetts Institute of Technology. He is a Distinguished Graduate of the Naval War College. RADM MacKinnon has extensive experience in the design, construction, engineering, and maintenance of ships and submarines, as well as in search and recovery operations at sea. He served on the Marine Board committee that assessed the National Oceanic and Atmospheric Administration's fleet replacement and modernization plan and is a member of the National Academy of Engineering.

Jacqueline Michel is a vice-president of Research Planning, Inc. She is a geochemist with extensive scientific and practical experience in the fate and effects of spilled oil on marine, aquatic, and terrestrial resources. Much of her experience is derived from work under contract to the National Oceanic and Atmospheric Administration as a member of the Scientific Support Team, which provides 24-hour emergency response support for oil and chemical incidents. Dr. Michel's areas of expertise include risk assessment and determination, technological recoveries, shoreline assessment, chemical countermeasures, and damage assessment to natural resources. She has authored several reports and scientific papers on the behavior, fate, and effects of Group V oils and has developed response options for tracking, containing, and recovering nonfloating oil spills. Dr. Michel is an adjunct professor of environmental sciences in the School of Environment, University of South Carolina. She received her B.S., M.S., and Ph.D. degrees in geology from the University of South Carolina.

R. Keith Michel, president of Herbert Engineering Corporation, has been with the company since 1973, working on design, specification development, and contract negotiations of container ships, bulk carriers, and tankers. Mr. Michel has served on industry advisory groups to the International Maritime Organization and the U.S. Coast Guard, developing guidelines for alternative tanker designs. He was a project engineer for the U.S. Coast Guard's report on oil outflow analysis for double-hull and hybrid tanker arrangements that were part of the U.S. Department of Transportation's technical report to Congress on the Oil Pollution Act of 1990. Mr. Michel served on the National Research Council Committee on the Oil Pollution Act of 1990: Implementation Review and is a member of the Marine Board. He holds a B.S. degree in naval architecture and marine engineering from Webb Institute of Naval Architecture.

James L. O'Brien has been the president and chief executive officer of O'Brien's Oil Pollution Service, Inc., since 1983. He is a former officer in the U.S. Coast Guard, where he was involved with pollution response, including an assignment as the leader of the Pacific Strike Team, a group responsible for responding to spills of oil and hazardous substances. He has been involved in responses to more than 150 significant oil spills, including well blowouts, vessel collision and strandings, facility releases, and pipeline ruptures, and he participated in spill-removal efforts during Desert Storm operations in Saudi Arabia. Mr. O'Brien's company provides services for companies and is the contract spill-management organization for a number of clients. He has published several articles in professional journals and has made presentations at national and international technical conferences.

Steven L. Palmer is a project manager in the Siting Coordination Office of the Department of Environmental Protection for the state of Florida. His professional experience includes aquatic ecosystem protection and watershed management assessments, air quality protection, and solid and hazardous waste management for both freshwater and marine environments. Mr. Palmer has testified as an expert witness in administrative hearings before legislative committees and at public hearings. He has served as a member of the U.S. Coast Guard Group V Oil Work Group Team, which examines issues surrounding the transport and cleanup of spills of heavy oils. He holds a B.A. in mathematics and marine science from the University of South Florida and an M.S. in civil engineering from Florida State University.

APPENDIX B

Participants in the Workshop and Meetings

COMMITTEE MEETINGS

First Committee Meeting, May 7–8, 1998
National Academy of Sciences, Washington, D.C.

Steven A. Anderson, Air Products, Inc.
Ken Bitting, U.S. Coast Guard
Louis "Coke" Coakley, Florida Power and Light Company
Deborah French, Applied Science Associates
Nelson Garcia-Tavel, Bitor America Corporation
Donna Leinwand, Knight-Ridder, Inc.
John Meehan, U.S. Coast Guard
Carolyn Raepple, Hopping Green Sams & Smith
Gloria Rains, Manasota–88
Jennifer Rains, law student, American University

Second Committee Meeting, August 20–21, 1998
National Academy of Sciences, Washington, D.C.

Charles A. (Andy) Miller, Environmental Protection Agency

Third Committee Meeting, October 15–16, 1999
Oakland, California

David Adams, Port of Oakland
José Dueñas, Greater Oakland International Trade Center
Peter Grautier, U.S. Coast Guard, Marine Safety Office
Jim Hardwick, Office of Spill Prevention and Response, California Department of Fish and Game
Harlan Henderson, U.S. Coast Guard, Marine Safety Office
Carl Jochums, Office of Spill Prevention and Response, California Department of Fish and Game
John W. Koster, U.S. Coast Guard, Chief Response Branch
Michael Latorre, Marine Spill Response Corporation
Douglas O'Donovan, Marine Spill Response Corporation
Steve Ricks, Clean Bay, Inc.
Scott Stolz, Hazardous Materials Response and Assessment Division, National Oceanic and Atmospheric Administration
Gail Thomas, Environmental Protection Agency
Edward Ueber, Gulf of the Farallones, Cordell Banks, and Monterey Bay (north) National Marine Sanctuaries

Workshop, August 20, 1998
National Academy of Sciences, Washington, D.C.

Ken Bitting
U.S. Coast Guard
Groton, Connecticut

Peter Bontadelli
California Office of Oil Spill Prevention and Response
California Department of Fish and Game
Sacramento, California

Kellyn Betts
Associate Editor
Environmental Science & Technology
Washington, D.C.

Duncan Brown
Marine Board Consultant
Arlington, Virginia

Barbara Davis
Oil Program Center
Environmental Protection Agency
Washington, D.C

Bryan Emond
Chief, Domestic Tank Vessel Branch
Washington, D.C.

Mervin F. Fingas
Environment Canada
Ottawa, Ontario

Nelson Garcia-Tavel
Bitor America Corporation
Boca Raton, Florida

Pamela Gibson
American Petroleum Institute
Washington, D.C.

Thomas Harrison
U.S. Coast Guard
Washington, D.C.

William Healy
Naval Sea Systems Command
Arlington, Virginia

Larry Hereth
U.S. Coast Guard
Washington, D.C.

Donald S. Jensen
Jensen & Associates
Elizabeth City, North Carolina

Eugene Johnson
Delaware Bay and River Cooperative, Inc.
Lewes, Delaware

James M. Kendell
U.S. Department of Energy
Washington, D.C.

Thomas Kniestedt
Apex Oil (retired)
Frohna, Missouri

Merritt Lane
Canal Barge Company, Inc.
New Orleans, Louisiana

John Latour
Canadian Coast Guard
Ottawa, Ontario

Steve Lehmann
National Oceanic and Atmospheric Administration
Boston, Massachusetts

Daniel Leubecker
Maritime Administration
Washington, D.C.

Malcolm MacKinnon, III, NAE
MSCL, Inc.
Alexandria, Virginia

Kathy Metcalf
Chamber of Shipping of America
Washington, D.C.

Mark Meza
U.S. Coast Guard
Washington, D.C.

Jacqueline Michel
Research Planning, Inc.
Columbia, South Carolina

R. Keith Michel
Herbert Engineering
San Francisco, California

Charles A. (Andy) Miller
Environmental Protection Agency
Research Triangle Park, North Carolina

Tosh Moller
International Tank Owners Pollution Fund
Houndsditch, London

James O'Brien
O'Brien's Oil Pollution Service, Inc.
Gretna, Louisiana

David Page
Bowdoin College
Brunswick, Maine

Steven Palmer
State of Florida, Department of Environmental Protection
Tallahassee, Florida

David Pascoe
Corbett & Holt
Washington, D.C.

Lita Proctor
Florida State University
Tallahassee, Florida

John Roberts
Coastal Towing
Houston, Texas

Gary Sergy
Environment Canada
Edmonton, Alberta

Gregory V. Sparkman
Division of Maritime Assistance Analysis
Washington, D.C.

Malcolm Spaulding
University of Rhode Island
Narragansett, Rhode Island

James Sweeney
Morania Oil Tanker Corporation
Stamford, Connecticut

John B. Torgen
Save the Bay
Providence, Rhode Island

Dave Usher
Spill Control Association of America
Detroit, Michigan

Glen Wiltshire
U.S. Coast Guard
Elizabeth City, North Carolina